TONGUE OF FLAME— PILLAR OF FIRE

At 7:17 A.M. local time, on 30 June 1908 in the Tunguska region of Central Siberia, an area of inaccessible forest, surging rivers, and roadless steppe, a tongue of flame shaped like a spear, a red fire-ball larger and more brilliant than the sun streaked across the sky followed by a train of dust and accompanied by a deafening roar, thunderous crashes and a terrific explosion. A huge pillar of flame shot upwards, cutting the sky in two. A black mushroom-shaped cloud rose into the atmosphere. People were thrown to the ground, windows broke, buildings creaked and groaned, the waters of rivers backed up against the flow, the driver of a train on the recently-completed trans-Siberian railway stopped his engine in the steppe, believing it had been derailed. A giant pressure wave swept twice around the Earth. The pendulums of seismographs agitated wildly. The following nights were marked by fiery glows and brilliant skies.

No one knew what had happened. To this day the question is unanswered.

We will send you a free catalog on request. Any titles not in your local book store can be purchased by mail. Send the price of the book plus 50¢ shipping charge to Leisure Books, P.O. Box 270, Norwalk, Connecticut 06852.

Titles currently in print are available for industrial and sales promotion at reduced rates. Address inquiries to Nordon Publications, Inc., Two Park Avenue, New York, New York 10016, Attention: Premium Sales Department.

THE TUNGUS EVENT

The Great Siberian Catastrophe of 1908

RUPERT FURNEAUX

LEISURE BOOKS • NEW YORK CITY

The most beautiful thing we can experience is the mysterious. It is the source of all true art and science.

—Albert Einstein
What I Believe, Forum, Oct. 1930

A LEISURE BOOK

Published by

Nordon Publications, Inc.
Two Park Avenue
New York, N.Y. 10016

Copyright © Rupert Furneaux MCMLXXVII

All rights reserved
Printed in the United States

Contents

1. Tongue of flame—Pillar of fire 7
2. Captain Cave inspects his barograph—Voznesensky reads his instruments 10
3. The fluttering of the wings of a frightened bird 17
4. The Flying Star with a Fiery Tail 29
5. Semenov shields his eyes, Kosotapov takes cover 37
6. Kulik explores 46
7. Kulik finds the cauldron 59
8. More expeditions and a fresh theory 71
9. Unique, unprecedented? 93
10. The investigators argue their theories 101
11. What happened? 115
12. Kasantev has the answer 120
13. Genetic mutation? 134
14. A Douglas fir tree in Arizona 141
15. Cosmological theories 150
16. But no exit? 162
17. Or, even more bizarre? 165
    Bibliography 171
    The Investigators 174
    About the Author 176

1

Tongue of flame - Pillar of fire

At 7.17 a.m. local time, on 30 June 1908 in the Tunguska region of Central Siberia, an area of inaccessible forest, surging rivers and roadless steppe, a tongue of flame shaped like a spear, a red fire-ball larger and more brilliant than the sun streaked across the sky followed by a train of dust and accompanied by a deafening roar, thunderous crashes and a terrific explosion. A huge pillar of flame shot upwards, cutting the sky in two. A black mushroom-shaped cloud rose into the atmosphere. People were thrown to the ground, windows broke, buildings creaked and groaned, the waters of rivers backed up against the flow, the driver of a train on the recently completed trans-Siberian railway stopped his engine in the steppe believing it had been derailed. A giant pressure wave swept twice around the Earth. The pendulums of seismographs agitated wildly. The following nights were marked by fiery glows and brilliant skies.

No one knew what had happened. To this day the question is hotly argued.

One only of the several theories which have been advanced would have been acceptable to the European and American scientists of that time. The other, more exotic solutions, would have been totally incomprehensible to people who had never heard of space-travellers, laser beams, nuclear fuel, anti-matter, gravitational collapse, black holes and white holes and hidden paths through super-space. Yet only three years earlier an obscure clerk in the Patent Office at Berne, Switzerland, had thrown open a window disclosing a new world of unimaginable and bewildering complexity. Albert Einstein's Special Theory of Relativity published in 1905 and his General Theory published in 1916 caused, eventually, J. B. S. Haldane to suspect that the universe is 'not only queerer than we imagine; it is queerer than we can imagine'.

The 'Tungus Event', as it has been laconically called, may be more bizarre than we can imagine. For more than 60 years it failed to attract the attention of the outside world except as a possibly unique and apparently easily-explained natural phenomenon. Then doubts crept in. The new knowledge gained from nuclear research and greater cosmic understanding created new theories, exotic solutions, the last so startling that its fearful implications have gone almost unnoticed outside the rarefied sphere of astro-physics—a fresh twist to the domesday syndrome.

2

Captain Cave inspects his barograph – Voznesensky reads his instruments

Captain Charles Cave lived at Petersfield, Hampshire, England. As befitted an English country gentleman, a retired soldier and a diligent member of the Royal Meteorological Society he possessed his own microbarograph, an instrument which detects and records disturbances of air pressure, the model invented by Drs Shaw and W. H. Dines in 1902. Coming downstairs for breakfast on the morning of 30 June 1908, Cave tapped his barometer and inspected his microbarograph. Between the hours of 5 and 6 a.m. it had registered a series of severe jolts. Four similar waves had been registered within a period of two minutes, followed by a large number of rapid oscillations of the pen. Cave placed the time of the disturbances at 5.23 a.m.

This air perturbation was also observed by

Doctor, later Sir, Napier Shaw, a Fellow of the Royal Meteorological Society who lived in the Kensington district of London. He noticed on his microbarograph 'the succession of four undulations, commencing with a range of about five thousandths of an inch, lasting about a quarter of an hour and then violently interrupted by a sudden, though short, explosive disturbance, which set up different and much faster oscillations for a similar interval'. He calculated the time also at 5.23 a.m.

Similar recordings were made at the Meteorological Office in Westminster (5.06 a.m.), at Shepherds Bush in western London, at Leighton Park School, Reading, further to the west (5.20 a.m.) and at the University of Cambridge (5.07 a.m.). Synchronization of these wave motions indicated that they had been received in England at the probable mean time of 5.15 a.m. They had travelled from north-east to south-west over a distance of 160 km (100 miles). Similar recordings were registered at Jena and Potsdam in Germany, and at places as far apart as Washington DC and Batavia (Djakarta) in Java, where they were less perceptible.

At Potsdam Astronomical Observatory, its Director Dr Wempe found on the barograph both direct and reverse waves, the first showing a sharp increase of pressure on the scale, and arriving by the shortest route of the great circle from the focus of the cause at 5.45 a.m. on 30 June, and the reverse wave which had circled the whole Earth, at 7.37 a.m. on 1 July. The difference between the two instants was 25.83 hours.

The English recordings were reported at the

meeting of the British Association held in Dublin that year. It was estimated, on the basis that the air waves had taken approximately 16 minutes to cover the distance between Cambridge and Petersfield, that the rate of travel corresponding with the first pressure wave had been 323 metres per second, that of the trough 318 metres per second. The short sharp shock had travelled at 308 metres per second. By comparison the air waves recorded in England in August 1883, resulting from the catastrophic eruption of Krakatoa, situated in the Sunda Strait between Java and Sumatra, had travelled at 314 metres per second. Nothing approaching their violence had been observed since Krakatoa's explosion, the greatest volcanic eruption in historical times.

No one bothered to explain the phenomena and they remained a mystery until 1929 when Captain Cave learned of the great aerial disturbance which had been experienced in Siberia in 1908.

In European Russia there had been only one contemporary report. A man named Peter Sukhodaeff had telegraphed from Kansk in the Tunguska region, a town on the trans-Siberian railway, reporting violent seismographic and microbarograph oscillations to the Central Seismic Commission at St Petersburg, the Imperial Capital. Sukhodaeff stated that the first shock had caused doors, windows and votive lamps to shake. Subterranean rumblings were heard. Five to seven minutes later came a second shock, accompanied by rumblings. It was followed by a further though less severe shock. The Commission dismissed Sukhodaeff's story as nonsense because Siberia was not an earthquake zone. It may have distrusted a report from other than the Director of the Kansk Observatory. He, Golunin, did not feel or hear anything at 7:17 a.m.—he was asleep.

A. V. Voznesensky, the Director of the Irkutsk Magnetic and Meteorological Observatory situated on Lake Baikal, was far more vigilant. Descending to the basement where the seismograph was housed, he found it oscillating violently.

Voznesensky watched the oscillations of the East-West pendulum. He noticed a three-fold wave-like distortion of the line of registration followed by a large impulse which set the pendulum in continuous movement. The duration of each wave was 2·2 minutes, their amplitude 1-2 minutes. The location of his instruments, the sophisticated Renold Balance housed in a hermetically sealed case and together with the less sensitive Miln recorder suspended on pillars in an underground basement building protected by double walls and entered through five doors, excluded any possibility of nearby jolts. The seismographs were registering earth shocks. He thought it strange that they lasted for one and a half hours. There were other puzzling features. The shocks seemed to have begun weakly and locally and continued in a series of jolts of unknown origin. He registered the time of the peak at 7.17 a.m. The stronger reaction of the East-West pendulum in comparison with that swinging from North to South indicated that the epicentre of profagation had been at right angles, and therefore to the north.

Similar shocks of varying intensity and air

pressure waves were recorded at different local times at 65 meteorological stations situated throughout central Siberia. Loud noises were heard over an area of 1,000,000 square km (386,100 square miles) and very loudly in the region lying between the Yenisei and Lena rivers and Lake Baikal. Due to slow postal communications these records took some time to reach Voznesensky. He sent out questionnaires requesting information and he read the stories printed in the local newspapers. From this stock of data he reached certain conclusions which he published in 1925.

Whether or not Voznesensky perceived the cause of the disturbances at the time, he and other people living in an even wider area experienced that night and the following nights extraordinary optical phenomena, unusually bright nights, those of 30 June and 1 July not being night at all. In northern Siberia massive, glowing, silvery clouds were seen and far to the south in the Caucasus mountains the night of 30 June was long remembered as the 'White Night'.

The Director of the Meteorological Bureau of the Caucasian Institute of Tiflis, Apostov, was amazed to find that at 9 p.m., when the sun glow should have died away, no artificial light was needed for his observations. It remained light all night, the glow brightening at midnight. On the following night the glow lasted until 10 p.m. and dawn came an hour earlier than usual. This phenomenon lasted for ten days and the night skies did not finally return to normal until the end of August.

Similar effects were observed on the Black Sea

where a man named Tikshina recalled many years later having asked his neighbours at 11.30 p.m., 'Why is it so light?' The student Polkanov was more observant. As befitting a future Academician he noted in his diary under the date 30 June, the appearance of an 'extraordinary and rare phenomenon'. It was still light at midnight and he was able to read small print at 1 a.m. The clouds were lit up by a yellowish-green light that sometimes changed to a rosy hue. He thought the phenomenon was very mysterious.

These bright nights were noticed by people living all over Russia, and throughout northern Europe. Several people believed they arose from the glare of a distant conflagration. Similar atmospheric effects had been observed following Krakatoa's cataclysm.

The Danish meteorologist Torvald Kul found the luminous nights so remarkable that he noted in his diary on 4 July: 'I should like to know whether during recent times a large meteorite has appeared in Denmark or elsewhere'.

In England abnormally bright nights occurred during 30 June and 1 July. In London it was possible to read a newspaper at midnight, and the following night was marked by a brightly coloured sunset. In Mill Isle on the west coast of Ireland, both nights were as bright as was usual in the extreme north of Scotland. So bright was the sky in Glasgow that only stars of the first and second magnitude could be seen. The brilliant luminosity enabled very good photographs to be taken at midnight in Stockholm, and a photographer at Dornock in Scotland took excellent pictures of the Cathedral with an exposure of only 90 seconds. An observer at Stald, near Goteborg, saw in the northern sky an hour after sunset an extraordinary bright light, magical and even imposing in that fair summer night. It lasted until 3 a.m. when the glow changed to red in the north-west and green towards the north-east. The brightness was sufficiently intense at Bordeaux, France, to enable people to read small print out of doors at 9.56 p.m.

These spectacular night effects were absent from the southern hemisphere and North America. A search made by Professor R. Wooley, the Astronomer Royal, through the logs of vessels of the British Fleet which had been sailing in the Atlantic between the Azores and Newfoundland disclosed nothing unusual, a point which may take peculiar significance later.

3

The fluttering of the wings of a frightened bird

In the Tunguska region the morning of 30 June dawned fine and clear, promising a warm and sunny day. Some cloud persisted in the north, the residue of the low-pressure system which was giving way to anti-cyclonic weather. Most people were up and about early engaged on their daily tasks, peasants ploughing the fields hoping to glean a quick harvest in that region of perma-frost, tanners washing wool and cleaning skins in rivers and streams, boatmen poling up rivers, miners washing for gold, merchants and traders bartering for furs and pelts, railmen shunting wagons and driving trains, postmasters, officials, bureaucrats, political exiles, women drawing water and preparing food, and the Evenki nomads herding their reindeer.

Few of these 'Tunguski' as they were then called

could speak Russian. Their forebears had lived in the taiga, the region of dense forests, cascading rivers and treacherous swamps, once the domain of the hairy mammoth, since time began. They are believed to be the oldest Siberian race, fur-trappers and wanderers who retain their ancient myths and legends. One of these folk memories related (as a European traveller learned before 1914) that in time past the Evenki had seen strange apparitions in the sky and believed they had been visited by what we would now call 'spacemen'. Before the day was out they would think the legends had been singularly confirmed. Their old stories of fireballs, dancing lights and glowing objects were scornfully rejected by the Soviet officials who talked with them in the 1920s. Such hallucinations smacked of visionary enthusiasm, akin to pernicious religious idolatry.

The Tunguska is a huge region 750,000 sq km (300,000 square miles) in extent lying between the Yenisei river on the west and the Lena on the east. Outside Russia little is known about it even today. The only detailed maps available outside the Soviet Union are those developed from photographs taken by American spy-planes and satellites, closely guarded secrets. Fortunately I possess an atlas printed in 1907 which locates the principal towns of Kansk, Kezhma, Ilimsk and Kirensk, and the trading station of Vanovara, and the rivers Angara and Podhamennaya or Stony Tunguska and their chief tributaries. Kirensk lies 772 km (480 miles) north-east of Irkutsk, Kansk 611 km (380 miles) to the north-west. Seven hundred and twenty-five kilometres (450 miles) separate Kirensk and Kansk.

Vanovara is 965 km (600 miles) north of Irkutsk.

The people of Irkutsk saw and heard nothing unusual that morning. They may have been distracted by the curious spectacle of three large motor cars passing through their town, the survivors of the six that had set out from New York on 12 February and which, passing through Vladivostock on 22 May, reached Moscow on 18 July. Two completed the journey to Paris, thereby establishing that these newfangled monsters were capable of traversing three continents.

'It', whatever it was, was probably too far to the east and too high in the sky at Irkutsk to distract attention from the new wonder. Audible and visible experiences were confined to the large circle centred on Vanovara.

Only one resident of Kansk, near the western rim of the circle, the tanner Sarychev, claimed to have heard anything remarkable. He was washing wool in the river when he heard a noise like the fluttering of the wings of a frightened bird. It was followed by a single sharp bang, so loud that it caused one of his workmen to fall into the river. It was succeeded by several more bangs which Sarychev likened to subterranean rumblings.

When he was questioned in the 1920s Sarychev recalled seeing a radiant body, roundish and about half the size of the sun, flashing across the sky. It was followed by a band-like trail. Its appearance was very brief. His statement that it disappeared 'behind the hill' indicates that the object traversed the sky to the east of Kansk.

People living in the outlying villages heard loud

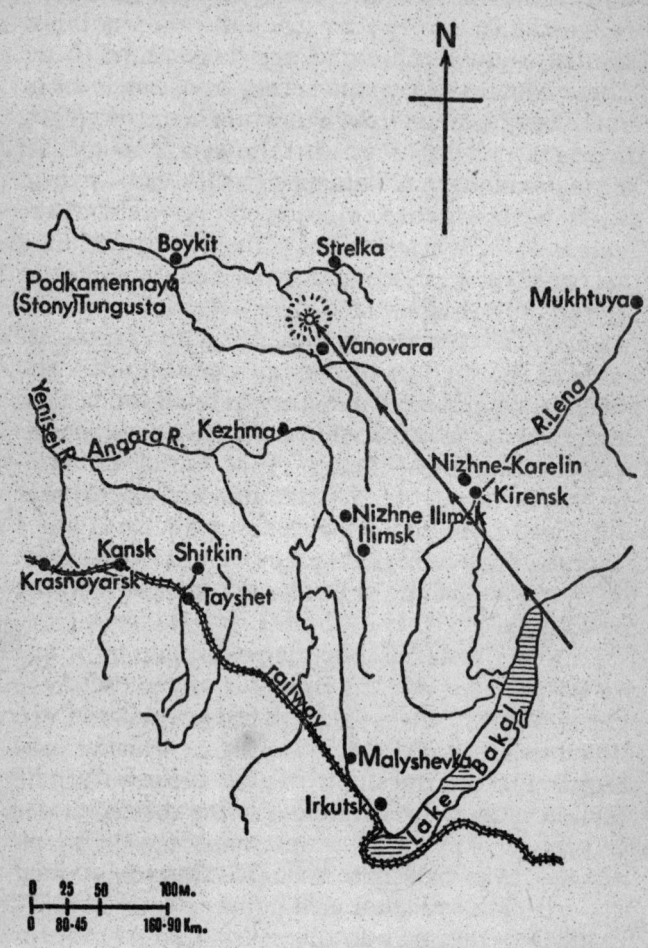

dull crashes like thunder followed by a prolonged roar and a second but fainter crash. Wooden house shutters rattled and china objects smashed to the floor. Animals became alarmed and restless. The villager Tropin thought that someone was rolling logs in the yard of his house. Going into the open he heard subterranean rumblings. A man named Vasilyev was puzzled because no earthquake followed the roar which he estimated to have lasted for two minutes. The air continued to vibrate for 15 minutes as though from a violent wind. Yet the day was cloudless and the sun shone brightly. The noises came from over the horizon to the north-east and ended with a fainter roar. Several villagers recalled hearing what they thought was gunfire to the north-east. Golunin, the Director of the Kansk Meteorological Station who did not himself see or hear anything, recorded these experiences. A man named Vronadzsky told him he had heard three short jolts, the first strong and the third weak, and then a loud roaring noise.

Golunin travelled along the railway collecting eye-witness reports from people at wayside stations. Denisenko at Ingash stated he had been sitting on a bench when 'suddenly a bang like thunder was heard'. The windows of his house shattered and he heard a second and third bang. They seemed closer and to the north-east. The loud rumble which succeeded these bangs seemed to be travelling along the surface of the ground.

Shayrev, the Director of the Troitskaya Meteorological Station situated on the rail line to the south of Kansk, reported hearing a noise which he likened to a shot fired from a large cannon, accompanied by a rumble and an echo. It continued for about ten minutes and was heard throughout the district. Sibirtsev at the village of Tina heard what he thought was the distant rumbling of thunder. Nemchenko at the village of Taseyevo was distracted from his ploughing by a thunderous roar for which he could not account because the day was bright and cloudless. A man in the village of Shitkina had a more vivid experience because the horses with which he was working fell to the ground from the shock which he attributed to thunder.

Thunder or an earthquake seemed obvious explanations to people living some distance from the area directly affected. Something had happened but no one had been hurt and no damage had resulted. When, eventually, they were told what had occurred, or what was thought to have occurred, they accepted that explanation because, as in the case of the European and American scientists, no other explanation seemed possible.

The Director of the Taishet Observatory dutifully completed the questionnaire sent to him from Irkutsk. All the buildings in the town had been shaken by a loud bang but there was no noticeable vibration; objects had fallen to the ground, doors banged and telegraph lines swayed. Witnesses living in remote villages heard thunder like cannon fire, weaker or stronger according to their locations. Some heard continuous rumblings and banging, others noticed separate bangs. The man Kaminsky believed that the thunderous roar lasted from five to seven minutes. He was told it had been caused by a train rolling over the wooden bridge, an explanation he refused to accept. Like most people he thought thunder had been the cause.

An educated witness, Dr Sergeyev at the village of Manzurka was busy tending patients at the hospital when he was startled by a noise like a shot from a high-calibre gun, something he may have personally experienced dressing wounds in the recent and disastrous war with the Japanese. The noise was accompanied by a slight wave-like vibration which ran along the ground as he determined from east to north. The peasant Pershin who had been working in the fields told him that the ground had given a slight jolt although the sky was absolutely cloudless.

A striking and even ludicrous incident occurred at Laika railway siding near Filimonovo Junction. The driver of train No 92 was so scared by the frightful roar and violent vibration of the air that he stopped his engine, fearing that it might be derailed. The passengers poured from the carriages to discover what had happened. When eventually he reached the junction he asked stationmaster Ilyinsky to inspect the wagons to see whether some of the goods they were carrying might have exploded. Another railman, the watchman Alekseyev, coming off duty reached his home to find the icon lamp swaying and the icon itself fallen to the floor spilling its oil. Unlike his neighbour, Shinelev, a canal watchman who heard a subterranean rumble, Alekseyev did not see or hear anything unusual. But in response to questions posed by the Director of the Meteorological station at Marituy he recalled that the icon lamp had been swaying from north-north-west to south-

south-east. Another of the many witnesses who answered the questionnaire, Kazansky, noted only a slight trembling of the air and sounds like thunder. The Postmaster of Khogot further down the rail-line towards Irkutsk heard three explosions like gunshots but noticed no air or ground tremor.

The reports gathered from the Kansk district, due to the town's location on the railway, reached Irkutsk earlier than those from other districts. They enabled Director Voznesensky to estimate that the 'place of the fall', as he termed it, lay 1000 km (620 miles) to the north-east of Irkutsk. Answers to his questionnaire trickled in more slowly from other areas, those not linked to Irkutsk by rail. We have no means of telling when they reached Voznesensky who had appointed himself co-ordinator of information about the phenomena, or how he assessed them at the time. He made no report to his superiors at St Petersburg who remained in ignorance of the nature of the phenomena or even that something remarkable had happened in remote Siberia.

Due to its connection by river with Lake Baikal news from Kirensk probably reached Irkutsk before that from more remote places which were also further from the path of the cause of the disturbance. The object passed through the sky some 130 km (80 miles) west of Kirensk.

The resident reporter at Kirensk of the Irkutsk newspaper *Sibir* supplied his editor with a detailed story in which he stated that: 'at the beginning of the ninth hour' – which indicates that he had no more idea of time than the lowliest peasant – 'a most unusual phenomenon of nature' was observed at the village of Nizhne-Karelinsk, 250 km (155 miles) to the north of Kirensk. High above the horizon to the north-west, the peasants saw a body shining too brightly to be watched with the naked eye, slanting downwards and disappearing within ten minutes. The people interviewed described the glowing body as cylindrical in shape and like a 'pipe'. They noted also a single dark cloud in the otherwise clear sky. The shining body fell to the ground where it seemed to be pulverized leaving in its place a cloud of black smoke. There came a loud crash which was not like thunder but resembled the noise caused by the fall of large stones or even gun-fire. Buildings shook and from the cloud sprang a forked tongue of flame. The villagers were panic stricken and ran into the streets. An old woman wept, crying that the end of the world had come.

The reporter had been in the forest a mile or so to the south of Kirensk where he heard what he assumed to be gun-fire for a period of ten to fifteen minutes. He hurried back to the town where he learned that north-west-facing windows had been shattered and that similar noises had been heard throughout the district. Frightened peasants told him of seeing a fast-moving red fire-ball streaking across the sky to the north-west, according to some horizontally, to others obliquely. One peasant farmer hotly denied it had fallen to the north-west stating stubbornly its course had been in the opposite direction. Everyone was talking about the strange phenomenon which they attributed to summer thunder, not unexpectedly for later in the morning the area had been soaked by heavy rain and hail.

The Director of the Kirensk Meteorological Observatory, Kulesh, got the time nearly right, reporting that between 7.15 and 8 a.m. that morning he had heard deafening bangs which he explained as like salvos being fired at the military station on the opposite bank of the river, a hasty rationalization which he qualified as 'an extraordinary natural phenomenon', when he noticed to his astonishment the vibrations on his barograph ribbon. This must have been placed in his bedroom because he added that 'during the time I had not got up and nobody came into the room'.

His curiosity aroused, Kulesh was quickly up and about collecting eye-witness reports. One resident told him he had seen at 7.15 a.m. toward the north-west a 'pillar of fire' about ten metres (9 ft) in diameter and in the 'shape of a spear'. As it disappeared he distinctly heard five loud bangs like cannon-fire one following the other. The sky where he had seen the fire was marked by a dense cloud. After an interval of about 15 minutes he heard more heavy noises. A boatman, an old soldier and a man of experience, backed up the townsman's story except that he had counted only 14 loud bangs. Far more people heard the bangs than saw the pillar of fire, possibly because they were indoors at that hour of the morning.

In his quest for information in the surrounding villages Kulesh was beset by bewildered people asking what had happened. Several, like the observant old soldier, had counted 14 bangs. Contractor Hanshin could not explain the behaviour of a plank which had fallen to the ground, because the air was perfectly still. Another villager had just started to open a box when he was rocked by a strong wind.

Later analysis of these reports indicated that the eye-witnesses around Kirensk had seen a fiery body followed by a dark cloud traverse the sky from south-east to north-west accompanied by loud noises and air perturbations which may have resulted from ballistic and explosive waves. It seems remarkable, however, that most people placed the aural phenomenon first in time whereas it is more probable that they had a visual experience first. No doubt the deafening intensity of the bangs had made the stronger impression. Estimates of the height of the object's trajectory varied between 80 km (50 miles) at its zenith to 10 km (6 miles) at its last observed altitude.

Other observers added to Voznesensky's growing fund of knowledge. They reported hearing loud bangs, windows shaking in their frames, trees bending as if in a gale and leaves quivering. But the technician Grechin, working at the water-gauge at Shamansk, reported no turbulence of the water.

Postmaster Klykov at Znamenskoe attributed the phenomena, which he aimed at 8 a.m. by his telegraphic operated clock, to an earthquake but he admitted to having seen a fiery streak in the sky which was difficult to reconcile with his theory.

Reports from even more outlying districts confirmed that hundreds of people had heard loud bangs and roars and had seen a conical fire-ball falling towards the north. Buildings were shaken and people and animals had been thrown to the ground. One observant witness described the fiery body as 'elongated and narrowing towards one end'. He called it a fiery dart torn from and as bright as the sun.

The town of Ilimsk situated about 160 km (100 miles) west of Kirensk lay less directly in the path of the object, but due to its location on the navigable Angara river it became a centre for the collection of information.

4

The flying star with a fiery tail

Polyuzhinsky, the Director of the Ilimsk Observatory, was way out in his estimate of the time when the phenomena occurred, stating in his report in Irkutsk that at 8.30 a.m. he had heard noises like thunder which became louder and louder as they approached. He described them as like shots fired from a revolver or the explosion of gunpowder. They ended with a terrible crash and an earth tremor. Hanging objects swayed and danced. He had seen nothing. Not so an inhabitant of the town who told him he had seen 'a flying star with a fiery tail'. It fell in the distance and its tail disappeared. Three women spoke of a 'ball of fire' which frightened them so much that they ran into the fields.

Two men, both inhabitants of Kezhma, were in a boat navigating the Angara river between that town and Ilimsk. Kokorin was steering and Bryukhanov was either a passenger or a member of the crew. They were on their way to the village of Khova to collect mill-stones. According to Kokorin, who related his story twice in 1921 and 1930, approaching the rapids he pulled the boat to the river bank making it fast at the village of Zaimskaya. Climbing the steep bank, he spotted a fiery red flame flying obliquely towards the earth. It was three times as large as the sun but not brighter. He watched the flame disappear behind the hills to the north-west and heard bangs which lasted for half an hour. The ground trembled and the windows of houses were shattered. But the water in the river remained calm. His crewmen were completely demoralized and it required considerable persuasion to make them return to their duties.

But Bryukhanov, on the contrary, declared he saw wedge-shaped rays in the sky while the boat was still some distance from the river bank. He must have been an observant man because he noted that the broad end of the wedge faced downwards. He watched the rays disappear behind the forest to the north. Interviewed in 1930 he indicated by holding up his hand that the rays appeared to be at the height of 60° above the horizon. He claimed that the water of the river had been disturbed. He joined Kokorin and with him entered a house just as loud explosions sounded which like his hosts he attributed to gunshots. He said that the fiery rays hurt his eyes and he had to look away.

The memory of Privalikhin, a boy aged fifteen at the time, had become so hazy by 1930 that he could not remember the day, month or year when the phenomena had occurred. But it was at the time of the harrowing of the land and on a morning bright and clear. He was in the fields and had just got one horse harnessed to the harrow and had started to attach another when suddenly he heard a sound like a single shot. He turned round and saw an elongated flaming object flying through the sky. Its front was much broader than its tail and its colour reminded him of a fire burning in daylight. It was many times bigger than the sun but dimmer so that he was able to watch it with the naked eye. Behind the flame trailed what seemed to be dust. It was wreathed in little puffs and the flames were followed by blue streamers. Its flight lasted for about three minutes before the flaming object disappeared over the crest of a hill to the north. He estimated its height at a little less than half the distance between the zenith and the horizon at summer sunset. He heard bangs like gunshots; the ground trembled and window panes were shattered as the object disappeared from view. He and other peasants ran into a hut in terror.

The villagers of Nizhne-Ilimsk were either peculiarly observant people or they may have been interviewed soon after the event. The reporter Ponomarev toured the villages near Ilimsk collecting eye-witness stories of the extraordinary occurrence. Correctly timing it at about 7.20 a.m. he noted that many inhabitants saw a fiery body shaped like a beam. It shot from the south towards the north-west before the people heard thunderous crashes. From the place where the body disappeared arose a tongue of fire followed by smoke. People living in the village of Karapchansk heard a terrible explosion. They ran into the streets when they heard more loud bangs. The vibrations caused houses to shake and glass windows to crack and fall out. One man spotted bright streamers high above the forest.

The postmaster at Nizhne-Ilimsk, Vakulin, naturally used the mails to record his impressions, writing to his chief that a number of villagers had seen a ball of fire crossing the horizon. It plunged downwards with huge velocity and a pall of smoke arose from the ground. About a minute later he heard the noise of a deafening explosion. Buildings rocked and shook.

Another villager, farmer Kokoulin, stirred to unusual activity by the postmaster's example, also put pen to paper. At 7.15 a.m. he and his labourers had seen a fiery object flying from south-east to north-west. Heavy bangs were followed by vibrations of the air. His people were terrified. He wrote again two months later, this time to Voznescnsky possibly replying to his questionnaire. The crash, he said, had been heard over a distance of 1500 km (932 miles). Another 'agriculturist', as the Soviet officials later pedantically described the man, a peasant owning the provocative and possibly dangerous surname of 'Romanoff', described seeing a ball of fire slanting across the sky and having at first a flattening shape. It appeared as two pillars of fire when it struck the ground. Romanoff heard several bangs and a noise like that caused by strong wind. Yet the day was completely cloudless.

The stories related and eventually collected from people living on the lower reaches of the Angara river, as it approached its confluence with the

Yenisei, varied in detail depending upon their distance and viewpoint from the centre of the disturbance. Everyone, wrote the editor of the newspaper at the mining town of Garilov, felt a trembling of the ground and heard loud crashes and bangs. The mine buildings creaked and groaned, the gold-washing machines quivered and the workers ran into the street. Horses fell to their knees and crockery cascaded from the shelves. The newspaper's correspondent in the village of Kezhma sent a long report describing a fearful crash which caused buildings to tremble. He thought the house had been struck by a heavy stone and he heard an underground roar which resembled a number of trains simultaneously clattering over rails. After an interval of five to six minutes he heard 50 or 60 more bangs which now sounded like artillery fire. As happened elsewhere, horses whinnied and fell to the ground, cows lowed and ran about wildly and windows shattered.

The anonymous reporter could not understand what had happened. The sky was clear, the waters of the river remained calm. Looking in the direction from where the sounds came he noticed a cloud of white ash clearly visible on the horizon. It grew smaller and more transparent and lasted until three o'clock in the afternoon by when it had disappeared completely. People whom he interviewed over a wide area told him that before hearing the bangs they had seen a fiery body fly across the sky at prodigious speed. Several eye-witnesses described seeing, as the flying object touched the horizon, a huge flame shoot into the air cutting the sky in two.

After the tongue of flame had disappeared they had seen a cloud of smoke and heard bangs. The glow, although it lasted for only one minute, was so strong that it reflected in rooms facing north.

One villager, gifted with a lurid imagination, and possibly an accomplished prophet of doom, believed that the earth was about to gape open and everything and everyone would be swallowed up in the abyss. Others, consumed by religious fervour, or victims of 'superstition' as the later Soviet chroniclers described them, were literally 'dumbfounded'. Hundreds of people described their experience on that remarkable day, men and women from all walks of life with the apparent strange exception of one calling, that of the Orthodox priest. Those who had survived the Red Terror of 1917–20 probably preferred that their one-time occupation should remain forgotten.

Voyaging up the Angara river, so to speak, we reach the village of Kezhma whose vigilant police officer immediately reported to the Provincial Governor the appearance of an 'aerolite' as he called it. The political exile, Naumenko, was far more explicit, being probably an educated man. He wrote 25 years later to the Soviet Academy of Sciences by which time his one time confinement in Siberia had made him a man of good standing. On the morning of 30 June Naumenko was working on a house with other carpenters. The day was unusually clear without a cloud in the sky. There was no wind and absolute silence reigned. Suddenly, very far off and scarcely audible, he heard what he took to be thunder. The sound became rapidly louder. A faint crash caused him to look at the sun. Its rays were crossed by a broad white band on one side, and on the other towards the north, by an irregularly shaped brilliant white elongated mass. It became many times broader as it advanced. It sped towards the forest and its disappearance was accompanied by another and far louder thunderous roar. Naumenko described it as taking the form of a shining ball, irregular in shape and far larger than the moon. His fellow workmen crossed themselves in stupefaction. Some of them fell backwards, stunned and terrified. He tried to calm them but they all abandoned work and fled into the village where crowds had gathered and everyone was talking about the phenomenon. As elsewhere, windows shattered and crockery fell off shelves.

Other residents were caught in their bath-houses. Forty-two-year-old Kokorin, another man of that name, had time only to remove his shirt when he heard sounds like gun-fire. He ran into the yard from where he saw, to the south-west, a red flying ball followed by rainbow-like bands. He watched it for three or four seconds before it disappeared towards the north. Like other observers Kokorin placed the sequence of events incorrectly, asserting that he had heard noises before he had seen anything, a natural mistake for the audible is more impressive than the visual. People do not habitually watch the sky and do not gaze upwards unless distracted by noise.

Another early riser had finished his bath before he heard anything. Half-dressed he ran into the street and looking up he espied radiant green, orange and red streamers, as broad as the street, passing over.

They disappeared and the bangs were repeated.

The peasant Bryukhanov was more advantageously placed ploughing his fields on a hillside to the west of Kezhma. He had sat down to eat his breakfast when he heard what he thought was gunfire an explanation which may have seemed feasible because he had heard 'talk of war', possibly some Czarist adventure in the lands to the south. A huge flame shot up behind the forest to the north and trees swayed impelled apparently by a hurricane. Bryukhanov leaped up and seized his plough fearing that it might be borne away by the whirlwind. His horse fell to the ground and the earth was whipped up. From the hill he saw a wall of water sweep up the river. Returning to his village he heard that windows had been blown out and people had been thrown to the ground. Horses became so frightened that they galloped off in panic dragging the ploughs behind them.

In the village of Nedokura the woman Kokinia heard a crash. Closed doors blew open and an old woman living with her fell off the stove (meaning probably the shelf above it, the most comfortable sleeping place in that cold climate).

These bewildered people living on the Angara river were less able to comprehend what they had seen and heard than those living further to the east and north where the object's trajectory was more directly overhead as it roared its way northwards making for Vanovara.

5

Semenov shields his eyes, Kosotapov takes cover

Vanovara is described as a 'trading station', a centre for the exchange of goods between the peasants and nomads and the merchants who supplied such basic necessities as sugar, flour, tea and simple tools. It comprised a few huts and could be reached from Kansk only by carts and pack-animals, the ubiquitous reindeer. Its tiny population may account for the scarcity of reports which were collected 15 and more years after the event.

By then Semenov had forgotten the year of the disaster but remembered that it had occurred when the fallow land was being ploughed preparatory to sowing. He had finished breakfast and had been sitting on the porch of his hut. He went into the yard and was just raising an axe to hoop a cask when suddenly 'the sky was split in two and its northern part was aflame'. He saw a great flash of light. A huge fireball filled the northern sky. He watched it for only a moment before he was forced to shield his eyes. When he looked again it had gone, and it had become dark. The heat became so intense that his shirt was almost burned off his back and his body was enveloped in flame. He struggled to pull off the garment and throw it away but a mighty blast threw him from his feet to a distance of seven feet or more. He saw clouds of earth split and rise from the patch of ground in front of his hut. A terrible noise shook the whole house and nearly rocked it from its foundations. The glass and framing shattered and the ground split apart. He lost consciousness and his wife dragged him on to the porch. He revived to hear a noise like stones falling from the sky and rattling upon the roof. He covered his head, afraid of being struck. The sky opened and from the north came a hot wind like the blast of a cannon. It made patterns in the soil, broke windows and tore off the iron hasp of a door.

Semenov remembered seeing his friend and relative Kosotapov who was working nearby collapse to the ground, seize his head in his hands and then rise and run into his hut. That morning Kosotapov had been getting ready for hay-making and needing a nail had been pulling one with a pair of pliers from a window frame which fortunately for him was situated on the south wall of the hut. Suddenly his ears were scorched by fierce heat. He clutched his head believing that the roof of the hut was on fire. Spotting Semenov prostrate on his porch he called, 'Did you see anything?' 'Of course I did', he replied, shouting that he had been overcome by the heat. Kosotapov went to his own hut but no sooner had he sat down on the floor than he was startled by an almighty crash and by earth sprinkling from the roof. The oven door flew off the stove and landed on the bed. A window pane burst inwards, scattering glass. Finally he heard a thunderous roar. He waited until all was quiet before going into the yard. He did not notice anything else, being screened and protected by his hut.

Semenov's nineteen-year-old daughter, Kosotapov's wife, completed the story. She and Martha Bryukhanova had gone to the spring to draw water. Suddenly she saw the sky to the north rent open to the ground and fire pour from the chasm. The sky closed again and the girls heard heavy explosions. They rushed away in terror, keeping their heads down, thinking they would be struck and forgetting their buckets. Reaching the huts, Semenov's daughter found her father lying semi-conscious on the porch and she helped carry him inside the hut. Twenty years later she vividly recalled the fire in the sky was brighter than the sun but she could not remember whether it also became hot. She had been too frightened to notice. But she thought the biggest bang had been directly overhead.

No wonder that these residents of Vanovara suffered from the radiant heat. Their village lay near the path of the incandescent mass and only 100 km (40 miles) from its blast. No wonder that Semenov at the time described his escape from death as a 'miracle', a term carefully censored from subsequent reports.

Later analysis of these stories suggested that Semenov and the Kosotapovs had seen in bright daylight a column of fire 20 km (12 miles) high and 1·6 km (a mile) wide, which became transformed into a mushroom-shaped cloud which soared to an altitude of 80 km (50 miles).

The fiery object left bewildered and frightened people in its wake. No one had been seriously injured although Semenov had had a nasty experience. It did not prevent him from making inquiries amongst the Evenki nomads who came to Vanovara to trade. But, gifted only with a few words of Russian, they were incapable of vivid description and much given to exaggeration as subsequent travellers discovered. Several were loath to speak of their experiences, others denied that anything unusual had happened and sought to conceal the location of the place where the fire had struck the ground. It had become a sacred place not to be divulged. The stories they told were often contradictory. They were recorded between 1921 and 1926 by three travellers, the geologist Sobolev, the engineer Obruchev and the ethnologist Suslov who luckily found 60 Evenki gathered together at the market at Strelka.

These Evenki, or Tunguski as they were then known, roamed the forest area to the north of Vanovara, the region drained by the Chambé river, with its several tributaries, the upper reaches of the Dilyushmo, Khushmo, Ilyuma, and Makirta streams. The Chambé is itself a tributary of the Khatanga or Podhamennaya, meaning the Stony Tunguska which joins the Teterga river near Vanovara. The fiery object, the Evenki stated, had flattened the forest in the basins of the Makirta and Khusmo streams.

Of these Evenki people Illya Potapovich proved the most loquacious although at the time he had been at Teterya, to the southwest of Vanovara, and 100 km (40 miles) from the scene of the holocaust. He had heard only long peals of thunder and felt the earth shaking. In 1921 he was able to interpret the experiences of his brother Ivan who was still too shocked to speak. Ivan, his wife Akulina and their friend Vasily Okhchen had been camping at the mouth of the Dilyushmo stream where it flowed into the Khushmo river, 25 miles south-east of the centre of the explosion. According to Illya, Ivan said that a terrible explosion had occurred and of so great a force that the forest had been flattened over a huge area, his hut had been blown to the ground and its roof carried off by the hurricane of wind. His reindeer had taken fright. He had been deafened and deprived of speech. The shock had brought on a long illness. When Sobolev interviewed the brothers, Ivan was still incoherent. When Illya turned to him to confirm the story, he became violently agitated, muttering in his own language, striking the poles of the tent, shaking it and gesticulating in his efforts to explain how his hut had been blown away.

By the time Obruchev reached the scene Ivan was dead. His widow related that while they were still asleep in the early hours of the morning their hut, which she called a tent, had been blown into the air, rising and flying away like a bird and they went with it. They fell back to the earth bruised and shaken. She and Ivan lost consciousness. When they recovered they heard thunderous noises and saw that the forest around them was ablaze. Many trees had fallen.

Vasily Okhchen said that when the hut was blown away he had been tossed to one side by the violent gust. The ground shook and he heard a terrible roaring. Everything was shrouded by smoke from the burning forest. The roar died away in time but the forest continued to burn. He set off with Akulina in search of his missing reindeer but many could not be traced.

Being asleep at the time these people saw nothing. Had they been awake they might have seen the flaming object plunge to earth or explode above ground, whichever it did. They might have had an experience not to be seen again until 16 July 1945 when the first nuclear device was exploded at Almogordo in the New Mexican desert. In fact, when the first pictures of the devastated Tunguska forest were published in the West after World War II an inquisitive and suspicious Briton asked his government to inquire whether an atomic explosion had not occurred on earth before 1945.

Suslov and the other early visitors to the district were forced to jumble together the often incoherent stories they gathered by questioning the nomad folk. They had all been awakened by loud crashes and explosions. The earth trembled and the air vibrated. The terrible air gusts threw them about and felled the trees. A great cloud, which they likened to the shape of a mushroom, rose in the sky billowing out as it ascended higher and higher. They declared that the explosion had burned trees, killed dogs and reindeer, destroyed their stores and their winter clothing. Several Evenki claimed they had found the charred carcasses of their reindeer. The number killed was not established, some said a thousand, others twenty. They said that 'the forest was crushed', 'the stone-houses were destroyed', 'the reindeer were annihilated', 'people were injured', 'the dogs were killed', 'the taiga was flattened', 'trees fell from the summit of a mountain', and the catastrophe 'brought with it a disease for the reindeer' causing scabs which had never appeared before the fire came. This disease from which the reindeer suffered became magnified in time to indicate the effects of nuclear radiation.

One Evenki told Suslov that the 'fire suddenly blew into the bank of the Chambé river', slightly below the mouth of the Khushmo stream where it quickly burned up two hundred reindeer belonging to Stephan Onkoul whose store-houses filled with sacks of flour and household goods were completely destroyed. His silverware and metal samovar had melted in the heat. He had found only the charred carcasses of his herd of 1500 reindeer.

Suslov endeavoured to learn exactly where the fire had struck the forest. Illya Potapovich said that, while out hunting squirrels, he had found, on the north eastern slopes of the Lakura Ridge close to the source of the Makirta river, a deep fissure ending in a large pit which had become overgrown with young trees. Andrei Onkoul confirmed the existence of this deep and wide pit in the area of the south swamp 'where the ground had twisted and turned and the water had gushed out', and which the Evenki people had earlier known nothing about.

Several offered to show him the flattened forest which they said could be reached in four days in summer by bark canoe or in three days in winter by reindeer. Suslov was unable to make the journey and contented himself by drawing a map based on their stories.

The Evenki agreed only that Ogda, the God of fire and thunder, had come down to earth that awesome day and burned people with his invisible fire.

Something like the Tunguska cataclysm may have happened before, giving rise to this belief. The ancient Finnish poem Kalevala (which may have derived from that migrant people's earlier sojourn in Siberia) describes a terrible fire in the sky which burned the forest, destroying life everywhere and causing rivers to form a great lake, possibly the reaction of the frozen tundra to an object crashing from the sky. The whole world became dark and even the abode of the Creator was without light until he created a new sun and moon.

Inquiries made amongst people living to the north of the disaster zone elicited little useful information. People at Boykit on the lower reaches of the Stony Tunguska, a hundred miles to the north of Vanovara, had seen a flash of fire to the south-east and heard loud bangs. In the village of Chadobetz a horse had torn the bridle from its rider's grasp. Pots and pans fell from shelves and windows shattered. In that area the fiery object was flying directly in the face of the sun which prevented it from being seen with the naked eye.

Mercifully the cataclysm and holocaust of fire

had occurred at a remote and virtually uninhabited spot. Only a few people had suffered from the searing heat and violent concussion. A huge flaming mass, a dazzling fire-ball, travelling at cosmic velocity, had streaked across the sky and plunged to earth exploding on or above ground, devastating, flattening and scorching an area of 250 sq km (80 sq miles) wrecking, it is claimed, eight million trees, setting up huge ballistic and seismic waves which travelled twice round the earth, leaving in its wake a gigantic pillar of cloud. The sound of the explosion was heard within a radius of 800 km (500 miles) and as far away as the Arctic Circle.

No one knew what had happened.

6

Kulik explores

The inhabitants of St Petersburg remained blissfully unaware that had the flaming object penetrated the earth's atmosphere four hours and forty-seven minutes later it would have pulverized their city, killing and maiming the bulk of the population. Their ignorance was complete despite Peter Sukhodaeff's swift telegram to the Central Seismic Commission. His report from Kansk, and the copy of the Tomsk newspaper which reached St Petersburg, aroused no interest. But someone copied the article's gist, telling the story of the derailment of the train at Filimonovo Junction and incorrectly stating that a large meteorite had been found buried in the ground near the railway line, on the back of a calendar published in 1910. The sheet was found in 1921 by Leonid Kulik, a geologist attached to the

Mineralogical Museum of the Soviet Academy of Sciences. Then aged 38, it excited his interest and launched him on a career which occupied the rest of his life.

But in 1921 Kulik had difficulty in persuading the People's Commissary for Public Instruction, Comrade A. V. Lunatcharsky, to authorize an expedition to search for the reported meteorite. That official's reluctance was not perhaps surprising for the young Russian State had only recently emerged from the Red Terror and the long war with the White Armies which followed the Revolution of 1917. Permission was finally granted and Kulik left Petrograd, as St Petersburg had then been temporarily redesignated before the adoption of the name Leningrad, on 5 September. He travelled by train to Kansk.

In the two reports he published in 1922, the first of which was misleadingly entitled 'The Lost Filimonovo Meteorite', Kulik left his readers in doubt whether he learned of the mass of data collected at Irkutsk by Director Voznesensky or read the Siberian newspapers which had published eye-witness stories in 1908. Voznesensky, who had yet to publish his findings, was convinced that the phenomena resulted from the passage of a large meteorite which had fallen far to the north of Irkutsk.

Kulik did learn that the report of a meteorite fall at Filimonovo was untrue. The supposed meteorite turned out to be a natural rock. He had come to Siberia to locate the place of the fall and collect fragments or possibly find the whole meteorite. Alighting from the train at Kansk he probably believed that he would return home in triumph, bearing positive proof of the cause of the aerial disturbance and pieces of extra-terrestrial matter for detailed analysis. His hopes were soon dashed. It took him eight years to reach the banks and tributaries of the Stony Tunguska, 700 km (435 miles) to the north-east of Kansk.

In 1921 Kulik learned that between 5 a.m. and 8 a.m. on the morning of 30 June 1908 a fiery object had passed over the Province and had fallen with a mighty crash somewhere to the north of the Stony Tunguska. The sound of its passage and explosion had been heard throughout central Siberia. Kulik returned to Petrograd where he reported the meagre results of his journey to the Society of Lovers of World Knowledge.

Voznesensky retired from the Directorship of the Irkutsk Magnetic and Meteorological Observatory, and published his report in 1925. He provided a map of the affected area on which he indicated the flight of the 'Bolide' from south-south-west to north-north-east, to its place of fall which had caused a forest fire. He concluded that the object had been a huge meteorite or group of meteorites which had fragmented 10–20 km (6–12 miles) above ground. He expressed his belief that future investigators who managed to reach that remote spot would find a large crater similar to the famous one in Arizona around which had been found a mass of meteorite fragments. He added that the Evenki nomads living in the taiga had a legend similar to the American Indians who still preserved the belief that their ancestors had seen a fiery chariot fall from the sky and penetrate the ground where the crater had been formed. The Evenki had stubbornly refused to show the place to the Russian investigators who asked to be taken to the spot. Voznesensky thought that scientific investigation at the place of the fall would prove a very profitable study.

Voznesensky sent the very considerable data he had collected to the Meteorite Section of the Mineralogical Museum of the Academy of Sciences where Kulik studied it. He also read the reports made by Sobolev, Obruchev and Suslov who had collected eye-witness stories.

Kulik was now far better informed than he had been in 1921. The phenomena had been far more remarkable than anyone had realized. For the first time within historic memory people had actually witnessed the descent and fall of a giant meteorite, if such it was. The event promised unique enlargement of human knowledge, especially on a subject about which little was then known. The Society of Lovers of World Knowledge and the Soviet Academy of Sciences agreed to send Kulik back to Siberia to find the meteorite or its fragments and to measure the huge crater it must have formed.

Kulik set forth again in February 1927 carrying the rough map Suslov had drawn from the descriptions of the Evenki people he had interviewed and a directive from Academician Vernadski urging the importance of finding the meteorite as quickly as possible in order to determine its dimensions, composition and structure. He told Kulik that his expedition was of great scientific importance and its results could pay its costs a hundredfold. In any case the money would not be spent in vain. Kulik was less certain. Privately he expressed doubt that he would find the meteorite. But he refrained from stressing his doubts in the diary he had started to keep and which he eventually published.

Accompanied by his assistant Gyulikh, Kulik left the train at Taishet, the next chief depot beyond Kansk, on 12 February. He needed to cross 700 km (435 miles) of country, covered by deep snow, intersected by frozen rivers and in intense cold. He collected stores, repacked equipment on the backs of horses, and set forth on 14 March making for the village of Dvorets on the Angara river. Acquiring additional supplies there he travelled along the river bank to Kezhma reaching that town on 19 March. He transferred his gear to three sledges drawn by reindeer, intending to follow the beaten track through the taiga but the packed snow on which the sledges slipped and slithered made progress very slow. The 200 km (125 miles) journey required three days of hard slogging. At last on 22 March he reached Vanovara, the little trading station where the Evenki came to sell their furs. He found Illya Potapovich there and recruited him as his guide. Illya, we recall, although not himself a close observer, had been able to interpret his shocked brother's story.

Next day Potapovich led Kulik and Gyulikh, both mounted on horses, in an attempt to penetrate the taiga and reach the area of devastated forest which he claimed to have seen. But the heavily laden horses were unable to force their way through the deep snow. The explorers returned to Vanovara where Kulik found gathered a number of Evenki who repeated the stories they had already told. Vasily Okhchen, who had been camped on the Chambé river with Ivan Potapovich and his wife, offered to act as an additional guide. Once again travelling on horseback the party, which included Illya, reached Okhchen's new hut on the Chambé as night fell on 8 April. Next day, joined now by Okhchen's wife and brother and having transferred their stores and equipment to reindeer, they plunged into the virgin forest following the rough track for two days until it ended blocked by impassable forest. They attempted to hack their way through using axes to fell trees. They were forced repeatedly to change direction and make detours. They reached and crossed the mouth of the Makirta stream where it entered the Chambé on 13 April.

Kulik, unfortunately, omitted from his diaries those colourful details which are now the explorer's stock-in-trade. No lugubrious howling of wolves disturbed his slumbers. Nor did the camp fire burn dangerously low untended by his sleeping companions. The pages of his diary remained uneaten by hungry reindeer or marauding bears and unsoiled by the slushy snow, for it was now springtime when the melting snow and cracking ice of the river made travel even harder.

But on 13 April Kulik could not restrain his excitement as, crossing the Makirta, he saw ahead uprooted trees, their tops pointing southwards just as they had been felled and stripped bare by the mighty whirlwind 19 years before. The direction of their fall indicated that the epicentre of the holocaust lay farther to the north up the Makirta stream among the snow-clad hills which Kulik glimpsed ahead. One hill on the Kltladni ridge, explained Okhchen, was called Shakrama Hill, meaning 'Sugar-Head', thereby describing its conical appearance, similar to the way in which sugar was then packed in cones. Two more ridges intervened before the centre of the devastated forest could be reached. Observing the steady increase of fallen trees Kulik led the way up the river.

At this point Okhchen refused to continue, saying he wished to return home, pleading insufficient provisions. More likely he was deterred by superstitious dread. He was refusing to carry out his obligations, Kulik angrily retorted. He succeeded only in delaying his guide's departure for two days, by handing over enough food for Okhchen, his wife and brother to reach their hut.

Guided now only by Illya Potapovich, Kulik and his assistant scaled the Kltladni ridge. From the summit of the sugar-cone hill they saw ahead a vast area of devastation which he estimated extended for 100 km (62 miles) from north to south and 40 km (25 miles) from east to west, where the fiery blast had scorched and flattened the forest leaving a vast windbreak, the debris of innumerable trees torn up by the roots. Their tops lay roughly parallel and pointing south. Kulik noticed isolated hilltops, their summits bare of growth, and little protected valleys where the fresh growth of trees was already 20 years old. The blackened forest contrasted sharply with the deep layers of snow and the grey background of the taiga.

He had not yet reached the epicentre of the cataclysm which was still obscured by more intervening ridges, bare patches denuded of trees. The evidence of his own eyes left Kulik in no doubt that a unique event had occurred, never before experienced within human memory. It could have been caused only by an explosion of unparalleled ferocity or by the impact of a body of gigantic proportions and impelled at cosmic velocity. It looked as though a giant had scythed down the trees.

Illya Potapovich pointed out the spot to the north-east where his brother Ivan and his wife and Okhchen had been camping, fortunately for them some distance from the epicentre of the holocaust, the 'Promised Land' Kulik could not now reach lacking sufficiently steadfast guides and all the paraphernalia of an organized expedition. Like Moses of old he had been accorded a glimpse of what lay ahead. His hopes to gain a wider view by scaling the hills to the north and south were frustrated by Illya's flat refusal to accompany him further. His superstitious fears of visiting the place where the 'heavenly body' had fallen overcame his sense of duty. Kulik could not go on alone for, lacking Illya's local knowledge, he and his assistant would be unable to regain Vanovara. Guided by Illya they returned to the trading post on 22 April.

Relaxing after his eight-day excursion into the taiga Kulik wrote up his diary, finding it hard to clarify his chaotic impressions. The whole majestic picture was too vast to be taken in. From where he had stood a mountainous region stretched away to the northern horizon, the distant hills covered with a white shroud of snow half a metre deep. Yet from his observation point no sign of forest remained. Every living tree had been struck down, scorched and burned. Around the edge of the dead area young trees were moving forward seeking sunshine and light. He thought it uncanny to see giant trees, 20 to 30 inches thick, snapped like twigs, their tops hurled hundreds of kilometres to the south-east and south-west. Only at the periphery of the area did the windbreak begin to merge with the verdure of the old taiga. Here and there, on summits and hilltops, the shattered trees showed up as white patches. Beyond, and as far as the eye could see, lay the endless, mighty taiga for which earthly fires and hurricane winds had no terror. Yet, so far, he had seen only the southern aspect of the burnt-out area.

Kulik determined before abandoning his search to reach and view the northern area of devastation, by travelling by raft and poling up the Chambé river and the Khushmo tributary above the Makirta. Recruiting four Evenki hunters he left Vanovara on 30 April and travelled by sledge across country to the banks of the Chambé. The ice had thawed and the intervening rivers were in spate, necessitating the unloading of the sledges and their portage by hand. Gyulikh, unhindered by baggage, followed another route accompanied by more Evenki who on reaching the Chambé built the rafts. They were ready by 9 May and the expedition set off, struggling against the flooding waters and floating ice. The raft-men had to steer their way, avoiding the ice-floes by pulling into the banks to allow the largest to pass. During one night the raft carrying the bulk of the provisions was torn from its moorings and swept downstream causing further delay while it was recovered.

Passing the mouth of the Makirta the explorers reached the Khushmo stream which was fortunately free from ice. One raft was left at its mouth with sufficient food for the return journey and another was built to take its place. The dry river banks enabled it to be towed by horses hired from Evenki hunters. Reaching the basins of the Khushmo and Chunya streams which ran westwards, Kulik climbed a hill from where the view enabled him to overlook the southern part of the devastated area. It stretched as far as the eye could see. Everywhere he saw flattened forest and burned trees.

The explorers travelled on for five more days establishing their 13th camp and encircling the range of hills which formed an amphitheatre enclosing the vast area he had already viewed from the ridge. The extent of the damage convinced Kulik of the radial nature of the devastation and he again noticed the direction in which the trees had fallen to the south of the central basin. He encircled part of this great cauldron, again observing that all the flattened tree tops, as though bewitched, faced south. A few remaining trees stood like gaunt telegraph poles stripped of their branches and foliage. As he descended the ridge it became dangerous to walk through the old dead forest as the wind rose, for twenty-year-old giants rotted at the roots crashed down on all sides. Some fell very close to him and their ever-present threat forced Kulik to keep his eyes on their swaying tops so that should they fall he would have time to leap aside. But with eyes raised on high there was the other danger that he would stumble on the poisonous adders awakening from their winter hibernation and more than usually dangerous in the spring, which abounded in the taiga.

Kulik reached the head of the Khushmo stream which convinced him that he had penetrated close to the centre of the fall, the place he described 'as a stream of water striking a flat surface splashes away in all directions', where the stream of hot gases and swarm of bodies had penetrated the earth and with explosive recoil had wreaked havoc.

The explorers had almost exhausted their provisions. They had sufficient food only for three days and the road back was long and arduous. Their expectations of living on game and fish had not been realized, although Kulik had shot a few duck. They had even shaken the flour bags and eaten the dust and a concoction made from the stems of young plants. Kulik's only thought was to get back safely and it was, in his own words, 'a question of flight'. They travelled for nine days by day and night down the Khushmo and Chambé rivers by when they had been reduced to plant food alone, keeping in reserve and estimating the weight of their horse. They reached the bank of the Stony Tunguska at the end of June just as the summer rains were beginning.

The result of his expedition, Kulik thought was, 'exceeding all the tales of the eye-witnesses and my wildest imagination'. The flattening of the forest could have been caused only by an air wave of tremendous power. The positive results of the expedition were irrefutable and, in his opinion, disposed of the sceptic's assertion, which had delayed the start of his investigation, that 'it was all pure fancy'.

Summing up his conclusions which he reported verbally to the Siberian Regional Executive Committee at Krasnoyarsk and in writing to the Presidium of the Academy of Sciences, Kulik expressed his thoughts thus.

The centre of the fall spread across the watershed of the Chunya and Khushmo streams forming a vast cauldron encircled by ridges, hills and summits and containing lakes, marshes and rivulets. The thousands, and perhaps millions, of flattened trees lay in roughly parallel rows, their scorched trunks denuded of branches, twigs and tops, their tops pointing away from the centre of the fall like a gigantic cartwheel. The flattened forest fanned out toward the rim of the cauldron. The former vegetation over an area of hundreds of kilometres bore the characteristic traces of uniform conflagration which was entirely different from the effects of an ordinary forest fire. Vestiges of bushes and moss grew on isolated islands amidst marsh and swamp. The marsh itself was pitted with dozens of peculiar holes varying in diameter from 10 m to 50 m (30 to 150 ft) and about 4 m (12 ft) in depth with steep sides and mossy base, its surface marked by a peculiar swathe caused, he thought, by the eruption of the original soil. Kulik was convinced that the meteorite had struck the marshy ground carving out a huge crater which, in the course of time, became refilled with water, thereby forming the swamp.

The holes, Kulik believed, (obstinately, thought some of his colleagues), had been made by meteorite fragments, the largest exceeding 130 tons in weight. He had been unable to explore them by digging. He immediately set about organizing another expedition which he hoped would excavate these holes and recover pieces of the meteorite and determine also the centre of the fall. The expedition would need to make a complete study of the area and spend several months at the spot. The Academy of Sciences once again agreed to supply the money.

The learned Academicians were unaware they were in danger of being scooped on the story of the disaster, for Harvey H. Nininger, the Curator of Meteorites at the Colorado Museum of Natural History, was trying to gain American support for an expedition to Siberia to investigate an event which he described as 'unparalleled'. He lamented not being able 'to secure what is yet available of this the greatest message from the depths of space that has ever reached our planet'.

7

Kulik finds the cauldron

Kulik left Leningrad again in early April 1928 determined to make a thorough examination of the country to the north of the Stony Tunguska river, the area of the fall. He was accompanied by the botanist Schoumiliva, the young scientist E. L. Krinov, the ciné photographer Strukov and the hunter and zoologist Vasily Sytin who would organize the expedition. Detraining again at Taishet the party, which included five other Russians, set off on sledges across country, forded the swollen rivers and reached Kezhma on 18 April. The 200 km (125 miles) journey to Vanovara took another seven days. There they learned that the rivers ahead were still blocked by ice and it was 25 May before they reached the banks of the Stony Tunguska. They travelled up the Chambé in boats towed by horses.

Strukov's film, and later Soviet pictures which have been edited to make the 25 minute film, 'In Search of the Tunguska Meteorite', (and which is available with a dubbed English commentary made by the Smithsonian Institute) demonstrates the enormous difficulties encountered and overcome by Kulik, the 'fearless and enduring man', as he has been called. Kulik and his helpers are shown trudging through deep snow, building boats and rafts, ascending the ice-choked rivers, climbing cliffs, building huts and examining the stricken area.

The explorers encountered a large waterfall and dangerous rapids which necessitated the unloading of the boats and the transportation of the equipment and stores by land. In the attempt to pass the empty boats through the rapids Kulik's boat was overturned throwing him into the turbulent water. Luckily his leg became caught in the mooring rope and he was able to drag himself to the bank, his spectacles intact on his nose. Strukov filmed the near tragic incident.

The explorers negotiated several more rapids and reached Kulik's old base camp within the basins of the Khushmo and Churinga streams on 6 June. They constructed a hut with a bath house and a separate store-room which they raised on stilts to protect their food from bears and wolves. They built a second hut on the ridge from where, in the previous year, Kulik had viewed the cauldron as he called the stricken area. Strukov photographed the fallen trees before he left to return to Leningrad. Sytin, the hunter, had the temerity to suggest that the devastation had been caused by a violent hurricane.

Eager as he was to locate the centre of the fall Kulik first determined its geographical position which he calculated at Latitude 60° 55″ N and Longitude 100° 57″ E, and determined its boundaries by placing marker posts to define the extent of the devastation.

He, or other investigators later, named certain features, the West Peat Bog to the north-west of his new base camp, Stolkovich Hill directly to its east, Farrington Hill further to the north-east and the South Marsh to the south-east. He pitched camp on solid ground between the Bog and the Marsh, and to the north of the Churinga (or Khushmo) stream, within the area he had named the 'cauldron' which was surrounded by the rim of low hills. Once again he noticed the curious radial fan-shaped pattern of the fallen trees, their tops pointing outwards and their roots towards a common centre. Yet several trees still stood gaunt and bare looking like telegraph poles, denuded of tops and branches.

Only then did he turn his attention to the holes and swathes in the South Marsh which had so excited him on his previous visit, when, lacking pumps, he had been unable to draw off the water with which they were filled. These holes or pits capped hummocks which rose amidst the Marsh.

The film shows that these hummocks were low and flat, some 15–100 km (30–50 ft) wide, and were surrounded by areas of soggy marsh through which the explorers waded wearing wooden plank-like 'skis', or summer 'snow-boots', to prevent their feet from sinking. Even so, they were forced to slog their way through the slush.

Kulik was delayed by the illness of two of his workmen, one of whom was suffering from boils due to vitamin deficiency. Sytin took him to Vanovara to recover and, on Kulik's orders, despatched a telegram to the Academy of Sciences requesting more money because the expedition was running out of funds. Kulik went to Vanovara to collect it on 4 August.

His return to the site was delayed by the illness of yet another workman. The man's condition deteriorated forcing Kulik to curtail his excavation of the holes. He cut a trench across the Marsh in hopes of draining them but still the water seeped in. He tested their contents with a magnometer but its sensitivity was too low to register any reaction. Although Kulik had shot a large elk, whose presence near the camp had been reported by the furious barking of the Eskimo dogs, food was running short and he returned to Vanovara where he had a fortunate meeting with the ethnologist Suslov who had collected the first Evenki stories. He went back with Kulik and helped to excavate one of the holes, which was 32 m (100 ft) in diameter and which in consequence became named the 'Suslov Hole'. It produced no meteorite fragments so Kulik returned to Leningrad. His failure to find any part of the meteorite raised doubts about the origin of the phenomenon.

Kulik was fully convinced, as he stressed to the scientists who assembled at the Mineralogical Museum to hear his report, that the cosmic block had lost matter and had begun to disintegrate as it entered the earth's atmosphere. Travelling at high velocity it had either split into molecules of matter, changing into a gaseous state, and splaying out and ejecting fragments, or had shattered above ground forming many small craters in the Marsh. The meteorite had made no marked crater because it had exploded above ground. The swathes in the Marsh, characteristic evidence of intense subsistence of water, resulted from a concentric surge caused by the pressure wave accompanying the object's flight. The Evenki nomads, he recalled, had testified that water had gushed from the earth where 'it' had fallen, somewhere in the northern part of the South Marsh. But he dismissed their stories that on visiting the scene, they had found pieces of brilliant iron. Kulik's obstinate insistence that he would still find fragments of the fallen meteorite failed to overcome his colleagues' doubts. The holes in the Marsh, it was said, were common features of Siberian marshes and had not been caused by pressure waves. Only extensive borings could substantiate his theory.

Kulik had failed to prove beyond reasonable doubt that the devastation of the forest had been caused by the fall of a giant meteorite but no other explanation then seemed possible. There was no large and deep crater. That was disappointing for the Americans had recently confirmed the meteorite origin of the huge crater in the Arizona desert.

Spurred on by the American achievement and anxious to allay his colleagues' scepticism, Kulik left Leningrad again on 24 February 1929, accompanied by E. L. Krinov, now the Secretary of the Academy Meteorite Committee who acted as deputy leader of the expedition and who would in time succeed to Kulik's role as the chief investigator of the phenomenon. The party included several specialists: the female Shumilova from Tomsk University, an authority on Siberian swamps; Afonsky, a drilling technician; five enthusiastic amateurs, Yankovsky, Optoutsev, Teminkov, Starovsky and Gredynha and the local workman Karamyshev. They brought with them two hand-operated bores, two swamp sounding drills, a theodolite, cameras and magnetic instruments and supplies they hoped would be enough to last for eighteen months. They travelled from Taishet to Vanovara in 50 horse-drawn carts.

Kulik immediately began excavating the pits in the swamp on the north-western rim of the cauldron which appeared to represent the centre of the explosion and he found another chain of holes in the south-eastern part of the swamp. The Suslov Hole, due to its position on the dry hillock, appeared to promise the quicker results. With immense labour the workmen and amateurs excavated a trench 38 m (124 ft) long and 4 m (13 ft) deep through the peat to draw off the water, carting away the still-frozen mud which they cut into blocks for examination by Shumilova who failed to trace any meteorite material. The discovery of a decayed tree trunk at the bottom of the hole dashed Kulik's hopes, for its presence disproved that the hole had been made by a meteorite which would have pulverized the tree. The absence of similar holes on the hillside where they might also be expected provided irrefutable evidence that the swamp holes were of natural origin. Krinov,

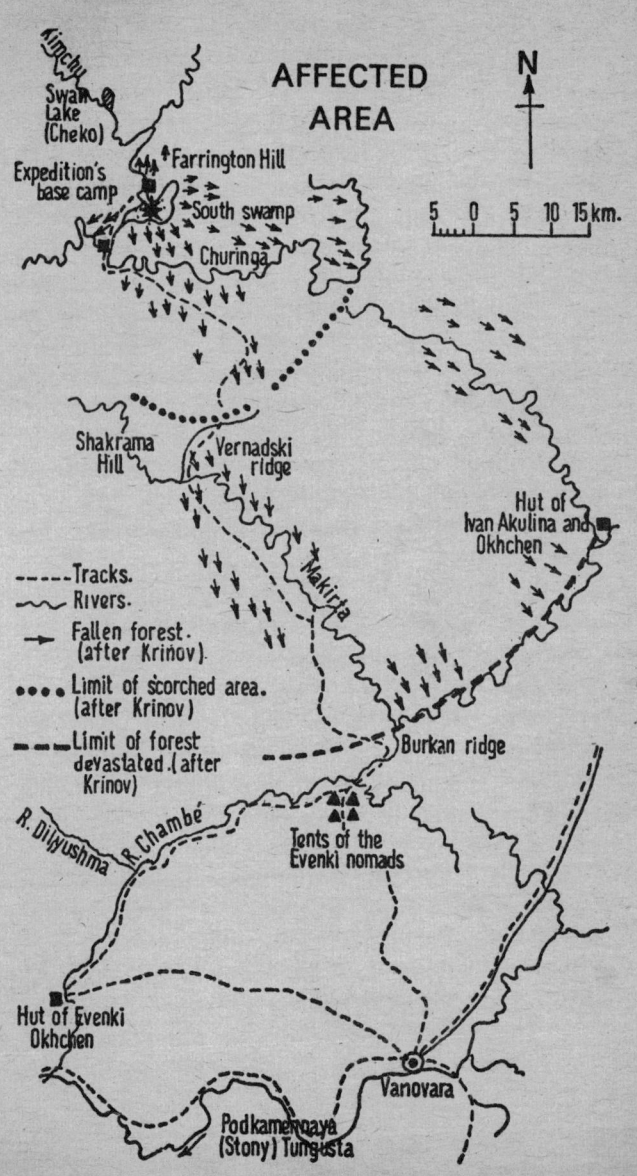

on the other hand, thought that the meteorite had gouged out a large crater in the swamp which had lost its shape and ceased to exist.

Kulik established a new base camp at the spot where he calculated, from the direction in which the fallen trees pointed, the meteorite had exploded or fragmented, whichever was the case. The vast area of devastation pointed south-east but many fallen trees also lay to the north, indicating that the explosion had radiated outwards in elliptical shape. He drew a map to indicate this and he added an arrow to show the meteorite's probable trajectory from south-east to north-west. According to Kulik's calculation the meteorite had passed 35 km (21 miles) to the north-east of Vanovara to explode above the southern part of the swamp. Astapovich, a later investigator, on the contrary decided that the meteorite's path had been about the same distance to the west of Vanovara. The question of the object's path through the sky will take greater significance later.

Kulik's calculations of the supplies required to keep the expedition at work proved less than efficient and he was forced to return to Vanovara. The party navigated the Chambé and Stony Tunguska in three large boats which due to the falling water had to be manhandled over shallows and shoals. The slow progress enabled Krinov to inspect the vast expanse of fallen trees, their tops slanting towards the south-east and their roots pointing towards the centre of the cauldron. Their configuration enabled him to observe that the radial nature of the devastation was far from uniform and that it bulged in the direction of the place where

Ivan, his wife Akulina and Okhchen had been camped on the fatal morning. Three of the amateurs left the expedition and Kulik replaced them with three other enthusiastic helpers.

The return of the party to the South Swamp was marred by the sudden illness of the workman Yankovsky who developed acute appendicitis. He was in delirium for several days before he could be taken by boat to Vanovara where he recovered. In Kulik's enforced absence, Krinov explored the swamp and located to its north the Swan (Cheko) Lake to which several of the Evenki had referred. Its position enabled him to delineate the northern zone of devastation 100 km (62 miles) from the centre of the South Swamp. He located the spot where Onkoul claimed that his grain-store had been destroyed. Krinov picked up a piece of a china cup and found ash from the Evenki's camp fire.

On his return and still wedded to the belief that the holes in the marsh had been caused by fragments of the meteorite, Kulik set his borers to work but soon realized the fruitlessness of the task. He roamed over a wide area collecting 89 soil samples which he placed in boxes carefully labelled. On his return home he put them in a drawer of his desk where they were found in 1957.

An injury to the one remaining horse and a scarcity of forage required yet another visit to Vanovara. It was by now November, ice was forming on the rivers, the ground was covered with deep snow and the thermometer registered 40° below zero. Several of the members of the party, trudging through the deep snow which soaked their boots and leggings, suffered from frost-bite and frozen feet. Krinov had to be carried by car to Kezhma. No gangrene had set in but he was forced to remain in hospital till mid-February by which time his big toe had been amputated. The expedition was in its habitual perilous financial state and was in debt to local traders for supplies. Kulik's pleadings brought additional money from the Academy of Sciences.

Anxious to rejoin the expedition, Krinov hired a reindeer drawn sledge to take him back to Vanovara through the deep snow. There he was introduced to Illya Potapovich who had overcome his superstitious fears and agreed to guide him on his return to camp. Illya indicated part of the swamp as the spot where, according to the Evenki, water had gushed from the earth at the time of the fall and pointed triumphantly to yet another hole as 'an undoubted meteorite fall'. But as far as Krinov could see the swamp was a treeless bog and fresh earth indicated the hole to be of recent origin. Krinov took Illya to Leningrad where he urged the necessity of an aerial survey to assess the area of devastation and determine its centre. But overcast weather that summer prevented aerial photography and when it was tried again in 1937 the aircraft loaned by the Polar Aviation Survey crashed in attempting to land near Vanovara. Its crew, Kulik and the photographer Petrov were unhurt.

Kulik, on his return home, lectured to a group of distinguished astronomers and geologists, illustrating his words with the film shot by Strukov. He described the stricken area of forest as taking oval form with the major axis lying from south-east to north-west. The trees lay in rows without branches or bark and pointing in the direction opposite to the centre of the fall. He had noticed this peculiar 'fan-shaped' pattern, but here and there some trees remained standing, denuded of branches and bark.

He painted a lurid picture of the effects of the explosion had it occurred above a highly populated area such as Central Belgium where no living creatures would have escaped death or serious injury. The central area of New York would have been blotted out instantaneously and glass in Philadelphia would have been shattered. It would have been one of the most appalling catastrophes in human history.

Kulik was now prepared to agree that the meteorite had exploded above ground causing the marsh to undulate and form the swathes he had seen. Far from being finished, work at the site was only just beginning. Somewhere beneath the semi-permanently frozen ground lay meteorite fragments weighing several hundred pounds.

Krinov found the absence of meteorite fragments puzzling, but he agreed that the Cauldron represented the 'place of the fall', as he described it. He found only one possible explanation of the contradiction. The meteorite had distintegrated high in the air above the South Swamp, scattering its fragments. He admitted to being perplexed by the blaze of heat which had scorched Semenov's skin and which seemed inconsistent with the fragmentation of a meteorite, however massive, and more consistent with the release of radiant energy.

Kulik revisited the scene in 1938 and 1939. His stubborn persistence in face of appalling difficulties won him the esteem of his colleagues who mourned his death on 14 April 1941 as a prisoner of war during the siege of Leningrad.

8

More expeditions and a fresh theory

Russians study of their own special terrestrial phenomena revived in 1950 by which time one at least of the more exotic theories of the cause of the catastrophe had been advanced, to orthodox dismay, and the original theory of meteorite origin had been questioned.

An extensive aerial survey of the devastated zone was undertaken, enabling a far more accurate assessment of the extent and inclination of the fallen timber to be made. The photomap confirmed that the blast had struck downwards from the height of several kilometres above the South Marsh, felling some 40,000 trees over a radius of 70 km (43 miles), considerably fewer, it would appear, than is stated in certain periodicals and books which exaggerate the total to '8 million trees over an area of 2000 square miles'.

The devastation was still clearly visible in 1954 when the renowned geo-chemist Kirill Florensky visited the scene, 46 years after the flaming object had exploded above the taiga, felling trees over an area he estimated at 3000 square km (1200 square miles). He confirmed the radial nature of the fallen timber and concluded that the so-called meteorite pits had no connection with the fall. They were merely cave-ins formed during the thaw of the perma-frost, a well understood phenomenon known as 'thermal sink-holes'.

Florensky returned in 1957 with Krinov and other specialists. They were able to fly from Kezhma to Vanovara, now a small town with an airstrip, hospital, school and radio-station, and penetrate the taiga by reindeer caravan.

They came again in 1958, the 50th anniversary of the phenomenon, bringing geologists, chemists, astronomers and physicists, and V. G. Fesenkov who had a special interest in the phenomenon for, as a young assistant in 1908 at the Tomsk Astronomical Observatory, he had been unable to carry out his observations for three nights because night had failed to fall. He would now become one of the leading authorities on the interpretation of the data. This expedition was the most ambitious so far. It disposed finally of the belief that the fallen meteorite had carved out a crater in the swamp which had become refilled. Probing at the centre of the swamp and its immediate vicinity proved conclusively that the 25 metre thick ice-layer had not been broken.

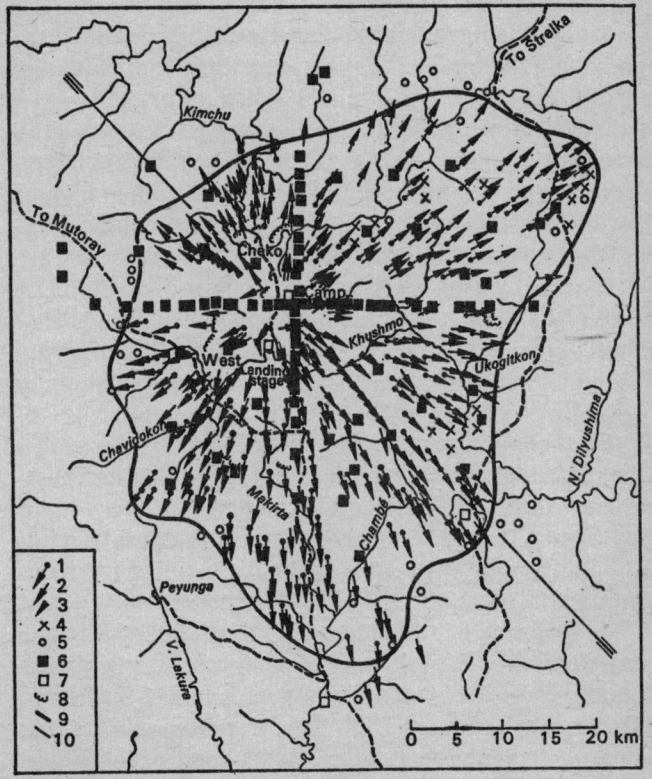

Map of forest destruction from 1961 data (Zotkin *et al.*). 1, 2, 3 – mean direction of flattened trees; 4 – places where fallen trees are observed with difficulty; 5 – places where there are no fallen trees; 6 – distribution of forest taxation areas; 7 – the expedition's huts; 8 – boundaries of the fire's advance as seen from the air; 9 – the general boundary of the area of forest destruction; 10 – south east variant of the meteorite trajectory; 11 – approximate boundary of physiological scorching of branches.

The team fanned out over a radius of 800 km (500 miles) searching for the crater which some members believed must have been formed somewhere and in hope of tracing meteorite ores. Three hundred soil samples failed to disclose them.

The chemists analysed soil samples taken from the area of the fall. They yielded a low concentration of magnetic and silicate dust which might have been derived from the explosion. Globules were found to contain 7 per cent nickel, 0·7 per cent cobalt and vestiges of copper and germanium, as well as iron. It was difficult to draw any definite conclusions about their origin because small cosmic particles are continually striking the earth from outer space, or they might have resulted from industrial contamination. Or they may have been caused by the hardening of dust content in the atmosphere. One tiny clue emerged. Many of these microscopic spherules of different composition had become fused into pairs, a feature found nowhere else on earth. This solidifcation could only have taken place at very high temperatures and from rapid condensation of incandescent gases–as might have resulted from the body's explosion.

The discovery of the samples collected by Kulik in 1930 and their spectral analysis by A. A. Yarvnel, a member of the Committee on Meteorites, suggested their similarity, both in chemical composition and shape, to substances found at the sites of large iron meteorite falls. The epicentre of the explosion, it was concluded, had been further to the south-east than had been previously thought, at the edge of the swamp rather than its centre.

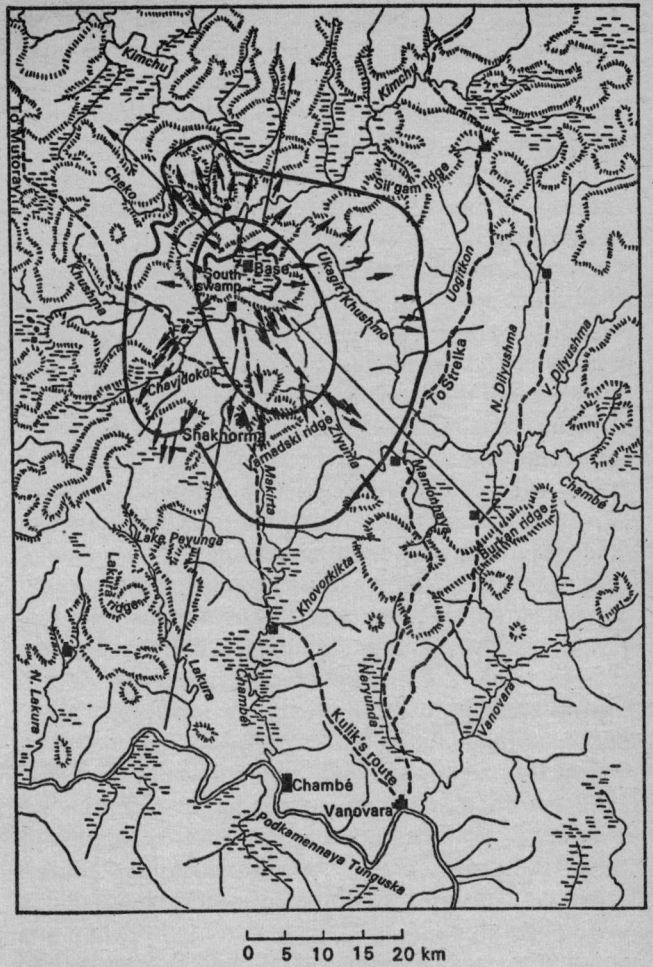

Map of the region where the Tunguska meteorite fell drawn up from the data of the 1958 meteorite expedition. Arrows indicate the directions in which the trees are uprooted; continuous lines – the limits of the devastation of different degrees; long arrows – the variants of the projection of the trajectory: A – from Astapovich, K – from Krinov.

It remained to analyse the results achieved by the expeditions since 1927. That was now made easier by far greater knowledge of meteorites in general and by the Russian investigation of the Sikhote-Alin meteorite which was seen on 12 February 1947 screaming through the sky above eastern Siberia and near the Chinese border. It disintegrated and fell in showers which pock-marked the ground, making 200 small craters, the largest being 26·5 m (30 yds) in diameter, and 6 m (20 ft) deep. Witnesses saw a dazzling white fire-ball moving from north to south, scattering a trail of fiery sparks which temporarily blinded eyes and cast shadows in spite of the brilliant morning sunshine. It left a trail of thick cloud and its passage was accompanied by loud rumblings resembling artillery fire and followed by roaring noises and crackling sounds like machine-gun fire. Military pilots stationed at a nearby base described the fire-ball as 'large as the moon', and equal in brilliance to the sun. The reddish-brown pillar of smoke stayed in the sky for several hours. The impact zone was spotted by the military pilots, the craters standing out like rusty-yellow blobs against the dazzling whiteness of the snow. The scientific expedition which reached the site within two weeks concluded that the single meteorite weighing 1000 tons had fragmented in the atmosphere. The Sikhote-Alin meteorite is now classified as an example of the impact variety which produce one or several small pits.

Large craters are formed by meteorites flying at high velocity which fall and shatter upon impact with the earth. The crater in the Arizona desert between the towns of Winslow and Flagstaff is the most famous example of this type of meteorite. It gouged out 300 billion kg (300 million tons) of rock in an instant of time forming a basin-shaped depression 1260 m (4200 ft) in diameter and 174 m (570 ft) deep, and causing a violent jolt which must have travelled around the Earth. Its mass has been variously estimated from 378 to 63,000 or even millions of tons and it is assumed to have been preceded by a wall of flame 20 million feet long. Its destructive force is believed to have been equivalent to that of a 20 megaton nuclear bomb. Extensive borings within the crater have so far failed to locate the residue of the meteorite. It is believed to have fallen some 25,000 to 50,000 years ago, with preference for the earlier date. That seems to dispose of the Indians' legend that their ancestors saw it fall because the first appearance of man in that part of North America is dated less than 25,000 years ago. But it is agreed that the Arizona crater conforms to the result of a nuclear detonation on the surface of the ground.

Nuclear or extra-terrestrial origin has not so far been attributed to the several other block-busters which have pock-marked the earth's surface. The one that fell between 5000 and 10,000 years ago on the Ungave peninsula in Northern Quebec Province, Canada, formed a circular lake 3341 m (10,960 ft) in diameter and 361 m (1180 ft) deep. It is named the Chubb crater after the prospector who found it. The Brent crater in Northern Ontario measures two miles in diameter, another at Odessa, Texas, is 168 m (185 yds) in diameter. Two craters in Australia, at

Boxhole and Wolf Creek, are 173 and 853 m (185 and 900 yds) wide. The Ashanti crater in Ghana, West Africa is six miles in diameter. The largest meteorite so far found, at Grootfontein in South West Africa, weighed 1000 kg (1 ton). In Greenland another weighed 33,000 kgs (33 tons), yet another found at Williamete, in Oregon in 1902, weighed 15,750 kgs (15·5 tons). All these craters are very ancient, as is primitive man's superstitious fear of meteorites which were believed to be hurled down by the wrathful gods.

Some meteorites have been venerated as religious objects, such as the Kaaba, the Black Stone of Mecca, and the 'Image that fell from Jupiter' which is linked by the *Acts* with the worship of the great goddess Diana at Ephesus. Several famous falls are recorded in history. A large meteorite in 616 BC broke several Chinese chariots and killed ten people. The Roman naturalist, Pliny the Elder, says that one as big as a cartwheel fell in Thrace in 476 BC.

A meteorite which explodes in the air with a loud noise is called a 'bolide'. It is the result of a rare combination of circumstances. It must be big enough to withstand substantial loss of material during its passage through the atmosphere. Its surface is heated by friction, leaving its interior cool. This range of temperatures sets up stresses which, in combination with its velocity, may cause it to explode releasing tremendous energy.

Large explosive meteorites are very rare. Most meteorites fall in showers providing magnificent displays, like the cascade of stones that fell over France on 13 September 1768. Examination of the fragments finally convinced French scientists of their heavenly origin. Previously meteorites were thought to have been of geological origin and still were as late as 1803 when US President Thomas Jefferson, on being told that two Yankee professors had reported the fall of 300 lbs of stones at Weston, Connecticut, thought it easier to believe that two Yankee professors would lie 'than that stones would fall from heaven'.

Meteorites originate within the solar system and are probably the debris of asteroids which themselves may be derived from the disintegration of a small planet many millions of years ago. They are of the same composition as the earth and provide evidence of the age of the solar system. Tests for radio-active decay performed in 1959 on a fossilized fragment found in North Dakota confirmed that it was 5 billion years old. Possibly even more exciting are the tests made on the meteorite which fell at Orgheil in France in 1864. Five 'well-organized elements', four containing aquatic algae, a primitive form of life, and the fifth entirely new, were reported in 1961. Isolated amino acids, one of the building blocks of life, were found in this fragment in 1972. But these discoveries are not universally accepted.

Doubts that the Tunguska phenomena had been caused by a meteorite had been expressed as early as 1930 by the meteorologist F. J. Whipple, the Director of Kew Observatory, who had collected the English microbarograph readings and had studied the lurid nightglows of 30 June and 1/2 July. His suggestion that the flaming object had been a comet, a 'dirty snow-ball' as he described it, was supported by the Soviet Academician Astapovich who had reached the same conclusion independently. He reasoned that a disturbance of such intensity could not have been caused by a meteorite because no meteorite, whatever its size, could have produced a cloud of dust which had risen so high in the atmosphere and had spread throughout northern Europe. Only the dust trail left by a comet could have caused the luminosity experienced in Britain, Sweden and France. It had reached ten times the normal nightly value. Astapovich thought also that the pressure wave accompanying a meteorite could not have travelled from Vanovara to England, a distance of 5720 km (3500 miles) within five hours, and he found in the Irkutsk records information which convinced him that the events of 1908 had been accompanied by perturbation of the Earth's magnetic field.

Astapovich's conclusions seemed to be confirmed in 1945 when Fesenkov learned that measurements made in California in 1908 established that there had been a noticeable decrease in the earth's atmospheric transparency from mid-July to mid-August, which could be explained only by the continuing presence of huge amounts of pulverized matter. This estimated loss of material, amounting to several million tons, represented more than 100 times the annual influx of meteorite matter. Only a comet's dusty trail could have so contaminated the earth's atmosphere.

Fesenkov inflicted another wound on the meteorite theory by pointing out that the object's trajectory had been in retrograde motion round the sun, thus overtaking the Earth's motion and reaching it head-on, an uncharacteristic and unlikely flight for a meteorite.

Nor would a meteorite, a product of an asteroidal distintegration, have had the large initial velocity required, whatever its mass, to cause the devastation which had been observed. And the absence of a crater and total lack of fragments ruled out the possibility that the object had been a meteorite.

On his return to Leningrad in 1958 Fesenkov expressed his belief 'beyond all doubt', that in 1908 a comet had blown up above the taiga, releasing enormous kinetic energy. It had been no ordinary comet. It had a mass of at least a few million tons. He relied upon two factors, the speed of the pressure wave recorded at Potsdam Geodetic Observatory directly and in reverse as it encircled the globe, at 318 km (200 miles) per second, and the rapid spread of cosmic particles through the atmosphere. This dust cloud, deriving from the comet's tail, had swept in a westerly direction away from the sun, following the same course as had the comet's nucleus up to the time of its disintegration.

Florensky agreed that the body was akin to a comet. As it descended rapidly through the atmosphere it lost mass and fragmented, became heated and finally disintegrated by a spontaneous process of deceleration and fragmentation.

The comet theory became established dogma.

Comets can be both large or small in size and may have long or short tails. Their appearance is rare.

Primitive and even mediaeval men accepted them as portents of evil, heralding war, plague, famine and the death of famous men. The appearance of Halley's Comet in 1066, which is depicted in the Bayeux Tapestry, was taken as a portent of the Norman Conquest of England.

Like meteorites, comets derive within the solar system but their origins are more uncertain. Careless wanderers, they pursue elliptical and sometimes erratic courses round the sun which may extend for millions of miles and endure for millions of years.

A comet's nucleus is composed of a gigantic block of ice, a loose collection of dust, frozen gases and rock, interspersed with water, methane and ammonia gases. It has been called a 'bag of nothing' yet it can release considerable power from the dissipation of kinetic energy and by chemical reactions. But this is theoretical for no comet, unless the Tunguska object was one, has exploded near enough to the earth for its composition to be assessed.

As the comet approaches the sun the gases liquefy forming a tail which may extend for millions of miles. The courses of some are predictable, such as Halley's Comet which appears in the sky every 76 years and was last observed in 1910, when its tail extended for 100 million miles. The Great Comet of 1843 had a tail reaching 300 million miles.

But in no known case, for example in 1066, 1861, 1882 and 1910, when a comet approached close enough for the earth to pass through its tail, has its cloud of dust produced luminous skies as did the object of 1908. The comet of 1910 had insufficient substance to cast a shadow. Even so, if a comet struck the earth in that year it was a unique event for no known comet has reached the earth or even approached it closely.

And why was not its collision course observed? The estimated dimensions of its nucleus at several hundred metres is only one degree of magnitude below that of well-known comets which were spotted at great distances from the earth. Its conspicuous tail could hardly have escaped notice as it neared the earth even in the early hours of daylight. The havoc caused shows it must have been a monster having a mass of several million tons.

Therein lies the paradox. Only the disintegration of a comet's tail, say the Soviet scientists, could have produced the dust cloud which permeated the atmosphere and spread across Europe, causing night glows and visual effects.

By comparison, the dust cloud raised by Krakatoa's great eruption of 1883 blackened the Sunda Strait, an area of 130 square miles, and, propelled by a south-easterly wind, spread out across the Indian Ocean forming an arc of twilight extending over 300,000 sq miles, and dropping its mass over 1,000,000 square miles. It encircled the earth covering 135,000,000 square miles, lingering for two years and causing vivid sunsets and brilliant night glows. Firemen stationed at New Haven, Connecticut, rushed out seeking a non-existent blaze.

And it may be a curious and possibly unacceptable coincidence that the earth was visited by a definite comet in September–October 1908. Moorehouse's Comet, as it was named, was watched clearly by Sir Arthur Eddington at Greenwich Observatory.

He estimated its velocity as 112,000 km (70,000 miles) an hour.

Neither this coincidence nor the anomaly of an unobserved monster approaching the Earth has shaken the beliefs of several leading Soviet scientists that the object was a comet. Others disagree, some violently. They are at one only in calling the event the greatest natural calamity known to man, a unique and unprecedented occurrence.

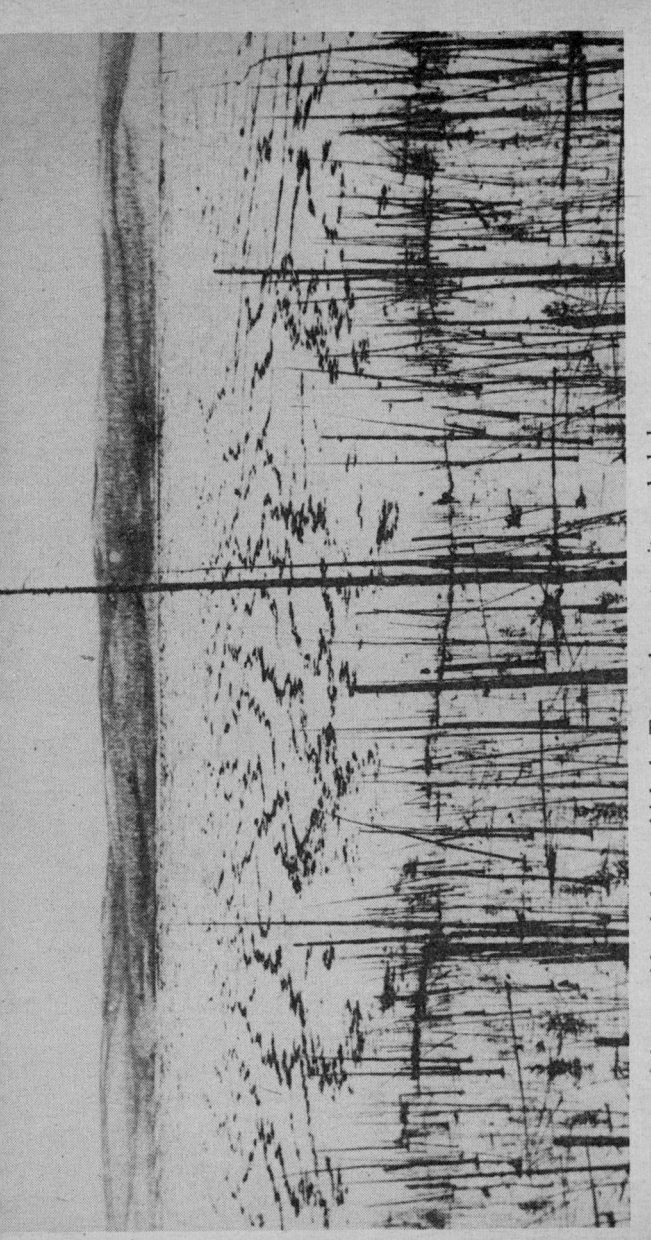

General view of the marshland above which the Tunguska meteorite exploded. (*Novosti Press Agency*)

Arizona meteorite

View of a section of the felled forest several kilometres from the centre of the Tunguska explosion. This photo was taken in 1930.
(*Novosti Press Agency*)

Opposite
The bark of a charred tree growing several kilometres from the epicentre of the Tunguska meteorite explosion.
(*Novosti Press Agency*)

Photo shows a cut of the trunk of a larch which was in the epicentre of the Tunguska meteorite explosion of June 30, 1908. The analysis of the annual rings confirms the hypothesis of scientists regarding the accelerated growth of trees in the area of the Tunguska catastrophe after 1908.
(*Novosti Press Agency*)

L A Kulik.
(*Pergamon Press*)

Witness Il'ya Popovich Petrov.
(*Pergamon Press*)

Witness S B Semenov.
(*Pergamon Press*)

After the explosion of the Tunguska meteorite a huge section of the taiga became a dead forest which you see on this photograph dating back to 1931. The houses in the background belonged to an expedition headed by L A Kulik, a prominent Soviet scientist and explorer of the Tunguska meteorite.
(*Novosti Press Agency*)

9

Unique, unprecedented?

Unique, unprecedented, perhaps, but whether the Tungus event is the greatest natural calamity experienced by man is arguable. Great storms, hurricanes, tornadoes, tidal waves have wreaked terrific havoc. Earthquakes have taken heavy toll of life. Volcanic eruptions have devastated towns and large areas of country. Their explosions ejected huge, uncontrolled energy, by the compression and sudden release of pressure.

The volcanic cone of Thera, now the island of Santorin, one of the Cycladic islands in the Eastern Mediterranean, following an initial outburst which submerged the island beneath huge layers of pumice, and upon the exhaustion of its magma chamber, blew its top in violent paroxysm, blasting off 83 square km (32 square miles) of its surface and creating a molten cavity into which the sea rushed. It was ejected with explosive violence forming a tsunami, a sea wave hundreds of feet high. It crossed 65 miles of sea and inundated the coast of Crete, swamping ports and villages. The ejected pumice descended on the fields rendering them infertile. Significantly, the Minoan civilization of Crete ended suddenly at that time, about 1475 BC, the date of Thera's explosion, giving rise possibly to the legend of Atlantis, the island state destroyed in a day and a night.

Far more is known about Krakatoa's great eruption. Long thought to be extinct, the three tiny cones rising from the little island in the Sunda Strait, between Java and Sumatra, resumed their volcanic cycle in August 1883. They ejected vast quantities of ash and pumice, which blanketed the islands, for two days, exhausting the magma chamber beneath the island. The final paroxysm began on 26 August. Drained of its supporting molten rock, the chamber collapsed. Into the superheated void poured the sea, to be ejected with explosive violence. The Krakatoa's Big Bang was heard 3000 miles away across the Indian Ocean. The tidal wave in this strait rose to 120 ft, drowning 30,000 people. It swept round the world, raising the level of the English channel two days later by two inches. The pressure wave encircled the earth seven and a half times and the dust cloud lingered in the atmosphere for two years. The island lost 4.3 cubic miles of its surface. The energy yield of the explosion is estimated to have been 10^{26} ergs, considerably larger than that of the Tunguska explosion, the nature of which has been discussed in a number of scientific papers. Their interpretation is difficult because much of the discussion and explanation is mathematical and cannot be easily translated in layman's terms. To render them meaningful, some understanding of scientific expressions is required.

Reference has already been made to the concept of 'energy' a term we may better understand by describing someone as an 'energetic person'. That means that he or she has a great capacity for doing work. This everyday use of the term is directly in line with the scientific concept, meaning the capacity to do work. Energy is found in many forms and it can be converted from one form to another in a variety of ways.

The automobile engine converts chemical energy into mechanical energy, with a great deal of that energy going to waste in the form of heat and noise. The method of starting a car using the self-starter illustrates the conversion of the chemical energy within the battery into electrical energy which, in turn, is converted into mechanical energy in the starter motor.

The prime consideration in this process of energy conversion is that energy is never lost. The total energy before conversion equals the total energy after conversion. This balance is called the 'law of conservation of energy' which means that the sum total of energy and mass can neither be created nor destroyed.

There are two basic forms of energy, potential and kinetic. Potential energy is the measure of the energy in any system, based upon the position of the material within that system. Thus, water stored behind a high dam has great potential energy due to the height at which the water is stored. Kinetic energy is the measure of energy arising from a body's motion. Various factors determine how much energy is contained in a moving body. To understand that a return to the word 'work' is required.

In everday life, the word work is applied to any form of activity that requires the exertion of muscular or mental effort. In physics, that branch of science which deals with fundamental concepts, work is defined more specifically:

Work is done only when a force is exerted on a body where the body at the same time moves in such a way that the force of some portion of it moves along the line of motion of the force's point of application on the body.

In other words, if a man stands still holding a weight in his outstretched hand, he may feel that he is expending a great deal of effort in preventing the weight from falling to the ground. But in the scientific sense he is doing no work as nothing is moving. If he now lifts the weight, he does an amount of work measured by a combination of the weight of the object and the distance he moves it. Thus, for the scientist, work is the combination of force moving through a distance.

Some textbooks define force as 'a tendency to produce change in the motion of a body upon which it acts'. Others elaborate the definition into a general relationship expressing force in terms of mass and acceleration. This carries us to the circular linking of concepts which presents for the layman the chief difficulty in understanding scientific work.

To understand the nature of a force, we need to understand the concept of mass. Mass is defined in terms of the ratio between force and acceleration, or the rate of change of velocity. The concept of acceleration is familiar to the car driver as he constantly experiences the relationship between the depression of the accelerator pedal and increase in speed. If we accept the concept of acceleration as the state of something being propelled at ever increasing speed, then we can dispose of the idea of mass by translating it into the more familiar concept of weight.

Isaac Newton in 1686 propounded the law of gravitation, with the help of the falling apple. It is stated thus: 'Every particle of matter in the universe attracts every particle with a force which is directly proportional to the product of the masses of the particles and inversely proportional to the square of the distances between them'. This gravitational attraction is what we experience as weight. Thus, we arrive at the result that all bodies, if allowed to fall freely, experience a force which accelerates them towards the centre of the earth at the rate of 980 cm (32 ft) per second, every second. This gives us a figure for the acceleration of a body under the influence of gravity.

If we consider the concept of mass as an expression of the amount of matter contained in any particular body, and measure it in grams, then a body of the mass of one gram, being accelerated under the influence of gravity, is being subjected to a force or weight of 980 dynes. The term dyne is used to describe a force which, if 980 units of it produces an acceleration of 980 centimetres per second every second upon a mass of one gram, then one dyne is that force which produces an acceleration of one centimetre per second every second upon that same mass of one gram. Thus the dyne is the basic unit of force within the centimetre, gram, second system of measurement. This lengthy search for a base unit to describe force illustrates the difficulty in trying to arrive at an understanding of scientific definitions without recourse to mathematics. But we still need to know what a dyne is in order to understand the measurements of energy used to compare the various estimates of the power of the Tunguska explosion. The dyne is a small unit.

So is the erg, the unit of measurement employed to express the energy released. It is a measure of energy derived from the Greek work meaning 'to work'. An erg is the amount of energy required to push one gram of matter for one centimetre. Earlier we defined work as being done when a force moves its point of application, through a distance. The units of force and distance required to define the erg are the dyne and the centimetre. Thus one erg is one dyne centimetre.

In most of the scientific papers dealing with the Tunguska phenomenon the energy, or work, yielded by the explosion is expressed as between 10^{23} and 10^{24} ergs. We find also expressions such as 7×10^{18} dynes. The difficulty with these expressions is to know how to read them. A cipher such as 10^{23} reads as 'ten to the power of twenty-three' and is merely a shorthand way of writing one figure followed by twenty-three noughts – a truly astronomical number. Figures such as 7×10^{18} dynes read as 7 followed by 18 noughts.

Another method accepted by mathematicians who want to convey that a figure lies midway between two possible extremes is to write, for example, (12·5±2·5) megatonnes. They mean that the figure is probably twelve and a half million tonnes, give or take two and a half million. In other words it can be 10 or 15 million tonnes.

We can form some idea of the magnitude of the Tunguska explosion by analogies more familiar than ergs and figures with long strings of noughts. Some of the scientific authors translate the energy figures in terms of nuclear explosions whereby an energy source of 10^{23} ergs has the same effect as the detonation of a 13 megaton bomb. One megaton has the explosive force of one million tons of TNT.

We may also be able to compare the power of the Tunguska explosion with the output of an electric power station. We are familiar with the term watt and readily relate it to such things as electric lights, heaters and all electrical equipment. The watt is the measure of the power consumed by such appliances. It is not confined to electrical equipment for it is a standard unit of power and is defined as the rate of working at one joule per second.

The joule is the unit of work within 'the metre' kilogram, second, system of measurement and as such as the 'big brother' to the erg in the centimetre, gram, second system. The relationship between joules and ergs, or between watts and ergs per second, enables us to translate the minuscule erg into the more tangible watt.

The watt is equivalent to 10^7 ergs per second. Whereas the watt is a handy expression for the power of small electrical appliances, the thousand times larger kilowatt is better employed for larger units. The output of electric power stations is rated in terms of megawatts, meaning one million watts. Assuming 10^{23} ergs (and larger figures are stated by some scientists) to have been the power output of the Tunguska explosion, we can try to express its output by considering the time duration during which this energy was yielded. Let us assume it was one second. Then the power output was 10^{23} ergs per second. One watt equals 10^7 ergs per second. Therefore by dividing 10^{23} by 10^7 we arrive at 10^{16} watts or 10^{10} megawatts, or, to set out the figures, ten thousand million megawatts. A small power station is rated at around 100 megawatts, a large one at 2000 megawatts. Thus the explosion while it lasted had a power output equivalent to one hundred million small power stations, or 10 million large ones.

We may be better able now to understand how the various investigators of the Tunguska phenomena reached their conclusions that powers of enormous magnitude were involved.

10

The investigators argue their theories

The scientific investigators approach the problem from different standpoints. One school of thought reasons from the physical evidence, the devastation within the stricken zone, the other draws conclusions from the seismographic recordings made at various stations.

V. G. Fesenkov, K. P. Florensky, colleagues I. T. Zotkin and M. A. Tsikulin, and A. V. Zolotov, derive their estimates from the observed effects. The Russian K. G. Ivanov and the Americans, Clyde Cowan, Willard Libby and C. R. Atluri, consider the seismic evidence, as does the Israeli scientist, Ari Ben Menahem.

Let us first consider the study of the parameters of the explosion made by Zotkin and Tsikulin in 1965. The term 'parameter' simply means the conditions and factors which need to be taken into consideration in the course of scientific investigation. As we have seen, speed and mass are the parameters to be reckoned with in calculating the kinetic energy of a moving body.

From the data assembled by the various expeditions to the site, Zotkin and Tsikulin drew a map of the stricken forest, plotting the outlines of the affected zone, which they assessed at 2200 sq km (850 square miles), also measuring the azimuths, or directions in which the 40,000 fallen trees lay. Then they constructed a model which could answer the questions of the size and height of the explosion and indicate its form. Was it an instantaneous 'bang', or did it derive from a rapidly approaching energy source in the form of a shock-wave?

This simulation of the Tunguska phenomenon is a classic simple piece of scientific experimentation. As they show by illustration, they made a scale model forest of plastic trees enclosed in a wooden frame from which they suspended an explosively charged cord, descending at an angle to the mini-forest and terminating at known height and position.

By this ingenious experiment they were able to simulate a variety of angles of approach and terminal heights. They could also vary the explosive charge and the characteristics of the cord. By repeating the experiment with varying parameters, they arrived at a resulting pattern of devastation which closely matched the observations of the damage at the site.

To fully appreciate the work of Zotkin and Tsikulin it is necessary to understand that they used an explosive cord, a device akin to a rapidly burning fuse. They did this to simulate the characteristic movement of a shock-wave front near the Earth's surface. But they were not content to simulate an instantaneous explosion at a particular point. The use of the moving explosive charge, they reasoned, would more closely simulate the approach of a body at high speed, pushing a great ballistic or shock-wave before it, the effect created by the tremendous disturbance caused by the passage through the atmosphere of a body at speed. To show that the energy released increased in power as the body approached the termination of its descent, they incorporated an additional explosive charge at the end of the cord to simulate the disintegration of the object. Thus, the explosive cord was designed to give out more energy as it approached the end of its run and final disintegration.

This combination of shock wave and final explosion satisfactorily simulated the observed devastation as the plastic trees became knocked down. From their experiment the two Russians concluded that the devastation had been chiefly caused by an approaching and descending air-pressure wave. It had energy of about 4×10^{23} ergs and terminated at the height of 10 km (6 miles) above the forest. The object was of a size and speed large enough to generate a ballistic wave of this energy.

Leaving for the moment consideration of the work of those Russian scientists who based their deductions upon the physical evidence, and due to their advocacy of particular theories of the nature of the object, we can consider the detailed work of K.

G. Ivanov who based his deductions in 1962 upon the seismographic recordings. It may be necessary to explain that the time lapse between their registration at distant points and the actual time of the explosion would be a guide to the height of the explosion on the basis of the known speed of travel of such waves.

Ivanov assumed the energy of the explosion had been 10^{23} ergs, the generally accepted figure. He made this assumption to provide a starting point for discussion, a typical example of scientific theorizing. The way in which he determined the height of the explosion is another example of neat scientific investigation. He recognized that the density of air varies with height and therefore there would be a difference in the time taken by earth-borne shockwaves to reach seismographic detectors. That would depend upon how close to the ground the disturbance occurred.

Let us follow Ivanov's reasoning, avoiding as far as possible mathematical details. First, he attempted to re-calculate the energy of the explosion, appreciating that such a calculation needed to take into account (1) the thickness of the atmosphere; (2) the density of the air at the point of the explosion; (3) a factor comparing how many times the energy concentration near the shock-wave front exceeds the average energy concentration within the rest of the explosion; and (4) the time taken for the upper boundary of the explosive shock-wave to reach a certain height in the atmosphere.

Having found numerical values for these factors, Ivanov estimated that the energy yield had been three to five times 10^{23} ergs, meaning that it was far greater than the figure arrived at by his Russian colleagues who based their calculations upon the physical evidence. He assumed that the discrepancy arose from the fact that they had used the air density existing at the Earth's surface, whereas it decreases at high altitudes. Ivanov calculated the air density at a given height above the Earth.

This calculation gave him the height of the explosion as between 8½ and 11½ km (5 to 7 miles) above the Earth's surface. The air density at this point made 10^{23} an appropriate quantity for the energy of the explosion. He was not content with mere coincidental verification of his original assumption which seemed to be belied by his own estimate. He searched for another method of checking his results, turning his attention to the time factor.

The formula for calculating explosive energy contains the factor 'T' which represents the measure of time taken for the boundary of the shock-wave to propagate from the point of the explosion to infinity. Ivanov found the value of 'T', in the case of the Tunguska explosion, to be 140 seconds. He determined this from the time difference between the beginning of the variations in the Earth's magnetic field, which is affected by large disturbances close to the earth, and the start of the earth tremors emanating from the region below the explosion.

These disturbances were recorded at a number of seismographic stations throughout Siberia and Europe. Voznesensky at Irkutsk (893 km – 555 miles from the epicentre) calculated the beginning of the earth tremors at 17 minutes past midnight GMT, and 7·17 local time, the corrected time allowing for the waves to have travelled from the point of propagation at 330 metres per second, the speed of sound.

This had been assumed to be the time of the explosion. Ivanov questioned this. Because the explosion occurred in mid-air it followed that the earth tremors began, not at the time of the explosion, but when the shock-waves reached the earth's surface. If the explosion had been high in the sky, it followed that the time taken for the shock-wave to reach the earth's surface would be correspondingly large. He thought it possible 'by determining this time interval, and by connecting by it the delay in the onset of the variations (at the seismograph)', to obtain a more accurate height for the explosion.

Ivanov selected a number of possible heights and calculated the time taken for the shock-wave from each height to reach the earth, comparing them with the delay in the onset of the recordings. From these comparisons he produced a corrected value for the time and thereby re-calculated the height of the explosion. This apparently circular method of producing data is a common scientific method whereby successive assumptions are made and compared with known data which is then reviewed in light of that comparison.

Ivanov concluded that the explosion took place at the height of 6 km (about 4 miles), but he qualified that statement by saying it may have occurred between 6 and 9 km (4 to 5½ miles). And he explains that 'if on the other hand the above method is used to estimate the height of the explosion on the assumption that the energy was two times 10^{23} then the height turns out to be between 3 and 6 km', the same figure as had been arrived at by Fesenkov.

Thus, Ivanov suggests values and offers an alternative to cover an energy source double that with which he had worked. If the energy was 10^{23} the height of the explosion was between 6 and 9 km, if twice 10^{23} it was 3 to 6 km. In other words, the larger the energy, the lower the height.

Ivanov's paper highlights the tentative nature of much scientific work. The willingness to doubt the evidence and question disputed reasoning accounts for much of what we consider to be scientific advance. It results from willingness to speculate. It would be possible but wearisome to evaluate the other Tunguska papers on this basis. Their discrepancies are remarkable and are summarized in the accompanying table.

It is advisable to point out that these authors did not investigate the same aspects of the Tunguska phenomenon. Some dealt with the size and speed of the object. Others did not. But most expressed opinions in the magnitude of the explosion and its height above the earth, and all were interested in its energy yield. Why should there be such a wide range of results? They estimate velocities ranging from a mere 5 km/second (11,000 miles per hour) to 60 km/second (about 700,000 miles per hour), and provide energy figures varying between 10^{21} and 10^{24} ergs. We find Cowan, Libby and Atluri reckoning

A COMPARISON OF THE DETAILS OF THE TUNGUSKA EXPLOSION.

NAME OF INVESTIGATOR DATE	MASS IN TONNES	DIAMETER IN METRES	VELOCITY IN KILOMETRES PER SECOND	EXPLOSION ENERGY IN ergs	EXPLOSION ENERGY IN MEGATONNES T.N.T.	OBSERVED DIRECTION OF FLIGHT PATH	HEIGHT OF EXPLOSION IN KILOMETRES	NATURE OF ENERGY RELEASE
FESENKOV 1961	10^6	233	60	—	—	South East to North West	5 to 6	Shock wave due to High Speed + Disintegration
BRONSHTEN 1961	2 to 7.5×10^4	—	16 to 30	—	—	—	—	—
IVANOV 1962	—	—	—	10^{23}	—	—	6 to 9	—
FLORENSKY 1962	—	—	—	10^{21} to 10^{23}	—	South East to North West	5 to 6	Shock wave due to High Speed + Disintegration
FESENKOV 1965	10^6	—	30 to 40	10^{23}	—	Ditto	10	Shock wave and chemical explosion
ZOTKIN AND TSIKULIN 1965	—	—	—	4×10^{23}	—	Ditto	—	Mainly due to shock wave
ZOLOTOV 1966	—	23	5 or 30	3 to 5×10^{23}	30	—	5 to 6	Thermal explosion either chemical or nuclear
COWAN, ATLURI AND LIBBY 1965	4 Anti-matter	1	60	10^{23} to 10^{24}	10 to 15	—	5 to 6	Chemical, nuclear or anti-matter
BEN MENAHEM 1965	—	—	—	3×10^{23}	—	South East to North West	8.5	Nuclear

the energy yield to have been equivalent to that of a 30 megaton bomb. But Ben Menahem is content with 12 mt.

These apparent discrepancies arise from the special scientific interest of each investigator. Scientists write in specialized periodicals for the benefit of their particular colleagues. Fesenkov, for example, as an astrophysicist, is concerned with the extra-terrestrial nature of the object, Ivanov with terrestrial magnetism. They are thus chiefly concerned to explain particular aspects of the Tunguska problem. The variety of answers explains their different emphasis. No one scientist has tackled the problem as a whole, a task awaiting accomplishment.

Writing in 1961 Fesenkov drew conclusions about the density, mass, diameter, velocity, direction and height of the body. He drew no conclusions about its explosive energy other than to say that, 'it must have been very great'. His particular emphasis is to explain his theory that the 1908 phenomenon was caused by some kind of comet or, as he guardedly put it, 'a body of cometary nature'.

Fesenkov's theories, which have already been briefly mentioned, can be summarized thus:
(1) The observed devastation of the Siberian forest would seem to indicate a very large expenditure of energy.
(2) The lack of evidence of any large body having actually struck the ground would seem to rule out the conventional meteorite which, travelling at relatively slow speed at impact, could not have

had sufficient kinetic energy to have caused the observed damage.

(3) He selects the phenomenon of the sky glows observed in Central Asia and North-West Europe, on the first night after the explosion, as the important factor in determining the nature of the body. The consensus of scientific opinion attributed this abnormal night sky glow to the penetration of some form of extra-terrestrial material into the Earth's atmosphere. The dust trail associated with a comet would be consistent with this theory.

(4) He makes a final point crucial to his argument for a body of cometary origin. Comets travel in retrograde (clockwise) motion round the sun, and opposite to the anti-clockwise motion of the Earth. Thus a collision between objects travelling in opposite directions is more devastating than if they are both going the same way.

Fesenkov remarks that the enormity of the destruction caused by the intrusion of an extra-terrestrial body is related to the speed at which it approaches the Earth, and argues that the degree of destruction is commensurate with the release of great energy. One of the principal ways in which a body can achieve such energy is to travel at great speed. Thus, if the Earth and the body could be shown to have been travelling in opposite directions, the questions of high speed and high energy would be answered.

To substantiate his hypothesis, Fesenkov reviewed the eye-witness stories which mostly relate that the body 'flew over them' from south-east to north-west, and points out that the apparent axis of the felled trees approximated to a south-east, north-west trajectory. At the time of the Tunguska disaster, the Earth's track along its orbit was directed in a north-south direction. Therefore, its collision with the body was almost head-on.

Fesenkov explains: 'We could interpret this by assuming that the body was either moving through space in the same direction as the Earth or that it was moving towards the earth'. He favoured the latter interpretation because it gave a higher relative velocity of encounter, a prime factor in calculating the energy released when two bodies collide. High speed would account for the high energy factor, and, as comets usually travel in retrograde motion, he concluded that it was the most likely body to fulfil the conditions associated with the Tunguska event. He rated the speed of its encounter with the Earth at 60 km (37 miles) per second.

But, how could a strange space-borne body, even one travelling at that speed, have penetrated so far and yet retain sufficient mass at disintegration to account for the devastation of the forest? Fesenkov estimated its mass at 'a few million tons'. No significant amount of material fell to earth and most of it was dissipated in the atmosphere, causing diminution of sunlight.

What effect would its passage through the atmosphere have had on it? Fesenkov assumed that his 'comet' had been formed of a compact mass with a density five times that of water, with a diameter of about 74 km (240 ft). He thinks it was a very dense, relatively small body which would have easily broken its way through the blanket of air in its path without it losing all its cosmic velocity.

His colleague, V. A. Bronshten, also considered the problem of the penetration of a massive body entering the Earth's atmosphere. Applying his thinking to the Tunguska problem, he proposed these characteristics for the body. It had a mass before entry of between 100,000 and 10,000,000 tons, and a velocity of 60 km (113 miles) per second. Its mass fell immediately prior to disintegration to between 20 and 75,000 tons, and its velocity diminished to between 16 and 30 km per second. These calculations implied that its energy was a whole order of magnitude larger than the more conventional 10^{23} ergs at entry, and about 10^{18} at disintegration. The body lost mass and velocity during its flight path when it radiated energy and heat in the form of shock-waves. It was to simulate just such a fast-moving approach that Zotkin and Tsikulin performed their experiment.

As befitting a careful scientific investigator, Fesenkov questioned his own hypothesis. He accepted that comets do not usually consist of a single compact mass of material. A density five times that of water was more consistent with the ironstone heart of a conventional meteorite. How then did his theory of the body's ability to retain a good proportion of its mass and velocity stand up when applied to a body having the low density usually associated with comets? Fred Whipple, he recalled, had described a comet's nucleus as a 'dirty snowball', made up of semi-solid and gaseous matter. Would not such a loosely constructed body simply break up into fragments on contact with the atmosphere? Fesenkov claims that it would not. The multitude of shockwaves would have interacted to form a common wave holding the particles together and enabling the body to move through the atmosphere as a single unit.

This view has been questioned by J. N. Hunt, R. Palmer and Sir William Penny who suggest in their article 'Atmospheric Waves Caused by Large Explosives' (*Phil. Trans. Roy. Soc.* 252,275 (1960), that at 70 km/second velocity the thrust on the nose of the object, in air density at sea level, punches a hole through it, and the thin ring-shaped residue would break up.

Fesenkov made estimates of the size of his comet, derived from two possible angles of trajectory, (1) a cosmic body with initial mass of 5 million tons and diameter of 738 m (½ mile); (2) a cosmic body with initial mass of one million tons and diameter of 233 m (765 ft). This question of size posed a problem with regard to density. His figures for mass and velocity indicated a density half that of water, even so a whole magnitude greater than that appropriate for a comet-like body. But the composition of its nucleus is conjectural. It may contain nothing but gaseous material or be a solid 'block of ice' with higher density than water.

Although Fesenkov makes a good case for his cometary theory, he is unable to offer convincing proof because serious doubts remain as to the ability of a loosely constructed body to penetrate so deeply into the earth's atmosphere without disintegrating.

But, as the disintegration is supposed to have taken place above 10 km, the body need not have travelled through the whole atmosphere, far less through its denser parts.

The examination of these scientific papers leaves many questions unanswered.

11

What happened?

What happened in remote Siberia on that beautiful June morning? The sun shone brilliantly in the cloudless sky, the wind blew gently from the southeast. It is 7.17 a.m. local time, 0.17 at Greenwich, England, about 7 p.m. the night before in New York.

A huge, flaming, incandescent, elongated spear-shaped object, resembling an enormous log or tube, with a gaseous aura, pale blue in colour, as or nearly as bright as the sun, a pillar of fire flying swiftly at great height, streaked across the sky and was seen and heard over distances 800 km (500 miles) from south to north and between places 500 km (310 miles) apart from east to west, affecting an area of 1 million sq km.

The object's passage was preceded by an intense shockwave, a sonic boom, which knocked down people and animals, shattered windows and doors, forced river water to surge up stream and frightened a train driver. It was accompanied by deafening noises which culminated in an ear-splitting mighty roar, and a vertical pillar of fire which split the sky in two as it exploded above the taiga, giving rise to a black mushroom-shaped cloud.

Had the object exploded above Chicago, the visual phenomena would have been seen at places as far apart as Pittsburgh, Nashville, Tennessee and Kansas City, and its noises heard in Washington DC, Atlanta, Georgia, Tulsa, Oklahoma and North Dakota.

It caused violent seismic and barographic shocks as its pressure wave encircled the earth in direct and reverse directions. Its pulverized dust swept across North Europe causing brilliant glows and, two weeks later, marked diminution of luminosity in far-away California. It scorched, burned and felled trees over an area variously estimated at 70 sq km (43 sq miles), causing widespread havoc which has otherwise been the exclusive domain of violent wind storms, volcanic eruptions and the detonation of nuclear and thermo-nuclear weapons. Yet it left no crater and no vestiges of material in its wake.

Taken unawares it shocked and frightened the people who glimpsed or heard it flying overhead. The Soviet investigators call their 'eye-witness' reports 'most trustworthy'.

At Malyshevka, to the north of Irkutsk, the object was seen to the east and flying in a northerly direction. From Kirensk and Nizhne-Karelin, to the north-east of Malyshevka, its motion was downward, towards the north-east and took the form of a tube. From the west, at Nizhne-Ilimsk, it was seen to the east. On the western rim of the visual and audible zone, at Kamenka, it 'broke away from the sun' which at 7.17 in the morning was 23° above the horizon. Its course was therefore somewhere between 30 and 40 above the meridian. At Kezhma, on the Angara river, it was observed flying high in the sky in a downward motion from south to north, and to the east of that place. It passed to the north-east of Vanovara and, consequently, reached the area above which it exploded from the south-east. The object's flight path is one of the few established facts of the occurrence.

The devastation caused is the sole surviving physical testimony. It is shown in the accompanying and self-explanatory diagrams drawn by the investigators who examined the stricken zone half a century or so after the catastrophe. The notes at the foot of each diagram represent the particular views of each individual. The diagrams show the roughly elliptical pattern of fallen trees radiating from a common centre. But the ellipse is not uniform for the havoc is accentuated in lobes. The longitudinal axis of the fallen timber extends in radius for 17 km (10 miles), the area of burning and scorching to 10 km (6 miles). Trees were felled within an area of 25 sq km (10 square miles) and uprooted, chiefly on high ground, to distances of 40–50 km (18–31 miles).

But not all the trees were felled. Two patches of scorched trees remained standing, looking like bare telegraph poles, within 17–18 km (about 10 miles) from the epicentre of the blast. They had been burned by an intense thermal heat flash which is believed to have reached the temperature of 5000 degrees Centigrade, only a thousand degrees less than the surface of the sun.

Analysing and attempting to interpret the evidence, a difficult task when the experts disagree, it seems that the object was first spotted flying at the height of 80 km (50 miles) when the friction from its passage through the atmosphere had caused it to heat up and become incandescent. At entry its mass may have been between five and ten million tons. It travelled at cosmic velocity, about 42 km (25 miles) per second, or at 90,000 miles an hour. It was preceded and possibly followed by a ballistic shock-wave and a vanishing cloud of dust, as it descended rapidly losing mass from friction. It may still have had huge mass with possibly a diameter of 100 m (over 100 yds). It exploded at an altitude of 10 km (6 miles) or possibly more or less.

The 'size' of the object can be compared with an inter-continental jet aircraft flying at 35,000 feet. From the ground it is visible as an inch-long body, tiny compared with the Tunguska body which appeared like an elongated tube or log. To the peasant imagination a log would have represented a known form of length, a massive object three or four feet long seen at that height.

What happened exactly when the body exploded is fraught with controversy. Zotkin and Tsikulin concluded from the action of their scale model that the ballistic pressure wave preceded the thermal heat generated by the explosion. This shock-wave, caused by the body's supersonic motion, became amplified by the violent break-up of the object. Descending both horizontally and vertically, the blastwave felled trees, knocking them down radially. The radiant heat from the explosion burned, scorched and blasted the trees within a narrower zone, creating a non-uniform effect entirely different from that deriving from an ordinary forest fire which would have consumed all the trees and their branches equally. Thus, the widespread devastation resulted from the intersection, intermingling and reversal of blast and explosive waves, descending in conical shape. The fire resulted from the sudden development of heat, and the deforestation occurred within two to three seconds. It continued to burn for days.

The amount of energy released has been estimated from the effects of the explosion. Whereas the Soviet investigators derived their estimates from observation, the Americans, Cowan, Libby and Atluri, in order to determine its power, fed the data, the distances at which trees had been blown down or ignited, into the Nuclear Bomb Effects Computer which yielded the value of 30 megatons, the almost universally agreed equivalent of 10^{24} ergs.

It is known that a nuclear explosion yields a million times more energy than any conventional form, whether it be chemical, like the expansion of gas, or mechanical, like the sudden release of a spring.

The nuclear theory, the first of the more exotic explanations for the Tunguska catastrophe, was first advanced in 1945, but in somewhat startling form for those far-off days.

12

Kasantev has the answer

In the summer of 1960 a party of young Russians, students and lecturers from several universities, or 'certain visionaries' as they were contemptuously called by members of the scientific establishment, paying their own expenses, spent several weeks camping at the site in search of traces of radio-activity resulting from the explosion of a nuclear propelled space-craft. They chopped down some of the trees that had been left standing, scoured the ground for cinders, inquired for survivors of the catastrophe and 'discovered' that several had died from an unknown illness which showed the same symptoms as derive from exposure to atomic radiation. Apparently they did not include Semenov and Kosotapov or the Evenkis who had been encamped within the disaster zone, all of whom appeared hale and hearty when they were interviewed in the 1920s.

The theory that a nuclear propelled space-craft accidentally exploded above the taiga in 1908 was first advanced by Alexander Kasantev, the well-known Soviet space-fiction writer in 1946 (*Vokrug-Sveta: Around the World*) and on his return from an inspection of the devastation caused at Hiroshima by the atom bomb. There is only one explanation, he declared in 1959 (*Visitor from the Cosmos*); for unknown reasons the nuclear fuel which propelled 'these secret visitors from the universe' exploded. The vehicle had broken up above the taiga. That accounted for the otherwise inexplicable absence of a crater. This is how Kasantev describes the disaster:

The explosion wave rushed down, and the trees directly below the point of the explosion remained standing, having lost only their crowns and branches. The wave burned the points of those breaks on the trees, and hit the perma-frost, splitting it. Underground waters, responding to the tremendous pressure of the blow, gushed up as those fountains seen by natives after the explosion. But where the explosion wave struck at an angle, trees were felled, in a fanlike pattern.

At the moment of the explosion, temperatures rose to tens of millions of degrees. Elements, even those not involved in the explosion directly, were vaporized; and, in part, carried into the upper strata of the atmosphere where, continuing their radio-active disintegration, they caused that luminescent air. In part, these elements fell to the ground as precipitation, with radio-active effects.

Kasantev's book was roughly handled by its scientific reviewers, V. V. Fedynskii and Yu G. Perel (*Soviet Astronomical Journal* 36, No 2, 1961). They found a 'consistent and conscious deception of the reader, in pursuit of one definite goal: to show that he alone, Kasantev, has discovered the true nature of the complex phenomena contrary to all the "conjectures" of the representatives of official science'. But they admitted that, 'it is impossible to say (and it is not maintained by anyone) that all the questions relating to the fall of the meteorite had been comprehensively answered'.

The Soviet aircraft designer, A. Yu Monotskov, supported Kasantev. Using electrical computers he reconstructed the flight of the 'thing' which he claimed had slowed down at the point of impact to a mere 0·7–1 km per second, the speed of a modern jet aircraft, whereas 10–60 km per sec (6–39 miles) would be the normal velocity for meteorites or comets. If the object had plunged downwards at this very low velocity, then, according to the laws of aero-dynamics, it would have needed a mass of 1,000 million tons to have caused the destruction which had been mapped and photographed. Its diameter could not have been less than 1 km (1093 yds) yet this monster had made no crater and left no debris.

It is claimed that the space-craft changed course abruptly at Kezhma, zoomed eastwards and turned back to reach the point where it exploded. The ambiguities of some of the eye-witness reports are thus employed to suggest that the object was intelligently guided.

Kasantev gained another supporter in the university lecturer F. Zigel who, writing in June 1959, asserted: 'At the present time, like it or not, A. N. Kasantev's hypothesis is the only realistic one in so far as it explains the absence of a meteorite crater and the explosion of a cosmic body in the air'.

Zigel concluded that the object could only have been an artificial craft flying from another planet, an explanation he slightly tempered by suggesting it might have been 'some extraordinary, still unfamiliar heavenly body'. His experience in training Soviet astronauts convinced him that the object had been steered because it followed precisely the re-entry corridor adopted by the astronauts who must navigate at an angle of minus 6·2 degrees to the horizon in order to survive without their vehicle burning up from friction by too steep a passage through the atmosphere or, alternatively, rebounding into space. The irregular shape of the stricken forest indicated that the explosion had fanned out elliptically. That was due to the explosive material being contained in a casing.

The well-known Soviet authority on aerodynamics who was responsible for calculating the courses of the Sputniks, Professor Sternfeld, gave his opinion that the space-ship had most likely come from Venus. Thus, according to Kasantev and his supporters, the object could not have been a meteorite because it made no crater. Therefore it must have been a space-ship. He conveniently ignores the acknowledged fact that it exploded high above ground. And what about the soil samples which were analysed in 1957 as containing nickel and cobalt? Kasantev has an easy explanation. They completely vindicated the space-ship theory. The metal and cobalt derived from the craft's steel hull, the traces of copper and germanium from its electrical instruments.

It is remarkable however that Kasantev first advanced his solution to the mystery in 1946, a year before the UFO controversy blossomed in the USA following Kenneth Arnold's picturesque description of the 'flying saucers' he had encountered in his aircraft above the Cascade Mountains in the State of Washington on 24 June 1947.

The UFO phenomenon is far older than that. Leaving aside the question whether the Biblical cities of Sodom and Gomorrah were destroyed by nuclear explosion, or Van Daniken's suggestion that the 'Gods' came to earth some 60,000 years ago to found a new race of men, or that terrestrial life sprang from the garbage left by careless explorers, the first modern sightings were reported in 1896 and 1897 by thousands of people living between Ohio and California. They saw 'air ships' in the sky several years before that form of aerial transportation became feasible. Thus they were accurately describing a form of aero-dynamics they had never seen or were likely to have imagined.

By a remarkable coincidence these sightings occurred in the year when H. Becquerl in Germany observed that uranium in minerals, without external excitation, emits radiation having a smaller wavelength than those of X-rays. It was also marked by J.

J. Thompson's detection of the free electron, two vital stages in the development of release of nuclear energy. These discoveries may have alerted a more advanced and observant Cosmic Society to take an interest, or to renew their interest, in terrestrial affairs.

Few scientists now deny the existence of life elsewhere in the universe or that many of these societies may have developed far farther than we have. Dr Carl Sagan, of the University of California, considers that the presence of intelligent beings in the universe is a mathematical probability and there may be frequent contact between communities with our galaxy, with space-ships shuttling back and forth between the stars. One or other may pay an occasional visit to earth in quest of information. While Sagan was still Professor of Astronomy at Harvard University he ventured his belief that there are in our skies 'objects which have not been identified'.

The UFO controversy is world-wide. All nations on earth have been 'visited' by forms of luminous phenomena which defy classification. The experts are divided. The scientific establishment is naturally wary of voicing opinions about anything as intangible as these countless sightings imply. Fearing to be branded as visionaries they dismiss these claims as hallucinations, hoaxes, even contrived frauds, the preserve of the lunatic fringe, the product of Cold War tensions, the threat of nuclear war and the dawning of the Space Age. Flying saucers, in their opinion, are just another example of irrational mass delusion. Others are willing to admit that 'something' is happening, a domain of nature unexplained by present scientific knowledge. It represents a fantastic challenge because nothing that intrigues the mind is ineligible for scientific study.

That is the view of the man who may be best qualified to discuss the problem–Dr Alan Hynek who acted as scientific adviser to the Condon Commission, the US Army Air Force's inquiry into the host of sightings which refused to 'go away'.

Hynek remarks in his book *The UFO Experience* (1972) that the craft seen in the skies follow 'non-random behaviour' and appear to be intelligently guided or programmed. They execute trajectories which no man-made aircraft could duplicate. They have the ability to hover and accelerate at high speed in the order of seconds without making sonic booms. In close encounter with terrestrial aircraft they follow a robot-like universal pattern making rapid descent and a few seconds later high-angled ascent until within a few hundred feet their bright luminosity vanishes. They occasionally change shape and produce durable physical effects on animals and inanimate matter. They are peculiarly localized and preferential in their manifestation and are observed at times and places when and where they are least likely to be detected.

Having subtracted obvious cases of hoaxes, mis-sightings of known aerial objects, and visionary hallucinations, there remains a profoundly disturbing body of data to gainsay, which is to accuse hosts of people from all walks of life in every part of the world of being crazy or lying. Many of the sightings have been made by people of at least average intelligence and sometimes embarrassingly above average. Hynek found himself driven to accept that the UFO phenomenon is something that is really new, whatever may be its nature. The British astronomer, Dr Percy Wilkins, agrees that many of the sightings cannot be dismissed as natural phenomena and that 'a residuum remains which cannot be thus explained'.

Our inability to explain the UFO phenomenon does not imply that the earth is being visited by intelligent beings from outer space. Unless they have discovered methods of travel nearly but not quite beyond our imagination, they would encounter the same difficulties as we now face in our hoped for explorations of space.

Numerous propellants are theoretically possible: nuclear energy either fission or fusion, nuclear particle emitters, electro-magnetic fields, photon rockets, solar heat (scooped up during flight), solar sail (the employment of solar energy by the same device as the principle of the handling of wind by a ship), and the use of the gravitational field of certain double stars by planet-hopping by a kind of cosmic sling shot.

It is conceivably possible that supplied with sufficient fuel, the quantity of which would need to be immense, a terrestrial space-ship could attain speeds close to that of light. Whither would it go, or from where may our visitors come?

Our galaxy, the Milky Way, contains more than 100 billion stars many of which must possess planetary systems. The qualities for life to develop are established by the example of earth. First the parent star needs to be about the mass of our sun. If it is too massive its gravitational force would slow the planet's rotation leaving one side always facing its sun. If it is too small the planet would be unable to gather sufficient heat. The planet itself must be suitably distant from the star, just the right size, right speed of rotation and right tipping of the axis, with a breathable atmosphere and a reasonable proportion of dry land. If it is large its gravity will crush everything on its surface. If small it would lose its atmosphere. About 17 million stars may possess planets having the correct mass. Even if only one in a hundred are orbited by a suitable planet it means that there would be 600 million habitable planets. If they have remained reasonably stable for thousands of millions of years, and are of correct chemical composition, life could have developed automatically as it did on Earth from organic components, the amino acids, the simple building blocks. There may be about 60,000 such habitable planets. But intelligent life may be rare.

The galaxy stretches in a flat arc for a 100 million light years and Earth lies on an out-flung arm 27 million light years from its centre. There may be 50 Earth-like worlds in our vicinity, 14 within 22 light years and a dozen as close as 14 light years. Proxima Centauri, the nearest star, 4½ light years distant, is thought to be unsuitable for planetary life.

For us, or for 'them', to travel the distance of 100 light years is not as simple as it sounds. That is due to the laws of relativity which contain a 'time dilation' or stretching factor, as Einstein perceived. According to Newtonian ideas of the universe it should take, for a ship travelling at the speed of light, ten years to reach a star ten light years distant and 20 years for the round trip. Einstein showed in 1905 that mass and energy are interchangeable.

Thus, if we accelerate an object it gains energy by virtue of its speed. It will therefore gain mass and will become increasingly harder to accelerate. As the speed of light is reached the mass increases sharply and the gains in speed become infinitesimal because all the additional energy is dissipated in increasing the mass. To accelerate the object beyond the speed of light would require burning up all the matter in the universe.

To achieve the speed of light or to travel faster than that absolute barrier is unattainable. At the speed of light the hands of the clock stand still. Beyond that speed they would reverse, going backwards in time which would violate the laws of causality for no consequence can occur before the event that precedes it.

A space vehicle and its crew travelling at even half the speed of light (670 million miles an hour) would experience curious effects. Common sense would suggest that the vehicle would be traveling at 1½ times the speed of light, at 1005 mph. But it does not, for the speed of light is constant and cannot be exceeded. In fact, time on the space-ship contracts, the clocks run slow and its crew and the ship itself shrink in size. As acceleration continues, they and it will become squashed to knife-edged thinness. But the crew, if they looked from their windows, would notice nothing unusual.

Instead of taking ten years to reach the terminal

star, ten light years distant, the space-craft will make the journey in five years. Yet to an observer watching from earth it will seem to take the full ten years. This is because the vessel's clock is running at half the speed of the clocks on Earth. Due to this time dilation effect the astronauts will reach a star 10,000 light years distant in ten years, one 30,000 light years away in 20 years, and for example, the Andromeda Nebula, 1,000,000 light years away, in 30 years. From there they would return to earth 60 years older to find that two million terrestrial years had elapsed.

Thus, unless the astronauts are willing to abandon their relatives and friends for ever and return maybe to a lifeless planet, space voyaging would seem to present insuperable difficulties by ordinary means of travel.

But these apparent difficulties do not rule out the possibility of space travel for, according to the latest theory which is described later, 'instant' timeless tunnels may connect the far reaches of space and even super-space. They may account for odd and not easily explained facts.

According to the Soviet mathematician Dr I. S. Shklovsky, an alien space vessel is parked within our planetary system. He suggests that Phobos, the inner satellite of Mars, is artificial. It is ten miles in diameter and revolves around the planet every eight hours, one-third of the time which Mars takes to turn on its axis, whereas no other moon in the solar system orbits its parent in less than a day, but all our artificial satellites do that. And Phobos is falling slowly on Mars at the known rate artificial satellites are dragged down by earth. Phobos is hollow, being a thousand times lighter than water. Is it a giant space-ship, a self-contained Ark parked in orbit while its crew explore elsewhere in smaller vehicles, and maybe are attracted to Earth by our enormous output of radio-energy?

The other 'fact' may be even more compelling. That is due to the massive research by Robert K. G. Temple (*The Sirius Mystery*, 1967). The Dogon, a small tribe inhabiting Saharan Africa, although they cannot see it, know that the invisible companion of the star Sirius, 8·7 light years distant from earth, rotates on its own axis, its orbital period is 'twice in a hundred years', it is immensely heavy and very small. Their knowledge was recorded by two French anthropologists between 1945 and 1950. The existence of Sirius's tiny companion has been known since 1844. It was observed in 1862 and photographed in 1970. It is a white dwarf, a collapsed star of incredible density and tiny size. It wobbles round Sirius in a regular 50-year orbit, thus orbiting its larger companion 'twice in a hundred years'.

How did the primitive Dogons acquire their amazing knowledge? They say that their ancestors who were not then in the Sahara were visited 6000 years ago by amphibious beings from the Sirius system. These beings may be pictured in the famous Tassili frescoes, rock paintings in the southern Sahara which appear to depict people wearing 'space helmets'.

Even if space travel is possible, whether or not Earth has been visited in the past, the Tunguska 'crashing spacecraft' theory suffers from a serious objection. A nuclear powered craft would be unlikely to explode. The characteristics of a nuclear explosion are described in the next chapter.

It may seem surprising that such a theory has been voiced in the Soviet Union where, according to Western ideas, free discussion is discouraged. On the contrary, where no Communist dogma is involved, sensationalism is encouraged both to show that Soviet science is free to debate issues and to provide an escape-valve from the rigidity of political censorship. Abominable snowmen are seen and encountered. Salamanders frozen solid for 5000 years come to life, monsters leave tracks in beds of oceans, ice falls from the sky and does not melt, mysterious forces invade the Earth in Russia as they do the world over. Belief in the supernatural seems to rise to a peak at times when human beings feel themselves oppressed physically or mentally.

In another variation of the Tunguska space-craft theory the catastrophe was caused by a too-powerful laser beam launched from a large planet circling the star Cygni 61. In our year 1894 the Cygnian scientists detected on their radio receivers an indecipherable message which came from Earth, 66 trillion miles distant. Believing they were receiving greetings, they attempted to answer. But for all their advanced knowledge they made two mistakes.

The signal was not a message. The radio pulses came from Krakatoa which had erupted violently eleven years before. Misunderstanding their source and believing that the waves emanated from an advanced civilization, the Cygnians despatched a laser beam in reply. When it struck the Earth's atmosphere its tremendous energy caused it to explode, becoming a giant bomb which blasted out a group of craters and knocked people down 30 miles away.

But the Tunguska explosion, whatever may have been its cause, made no craters.

13

Genetic mutation?

The blinding flash and terrible heat which melted Okhchen's silverware and scorched Semenov's skin, the brilliant fireball, the now familiar mushroom-shaped cloud, the intensely hot and luminous shock wave, the seismic disturbances, the cloud of debris and the extent of the damage, and above all the huge release of energy concentration into very small volume, seemed characteristic of the spontaneous release of nuclear energy.

This suggestion, advanced in 1960 by Soviet scientists, apparently overcame the difficulty of explaining the phenomena in other terms, for the energy released by nuclear weapons is a million times more powerful than the yield from more conventional forms of explosion. Had the Tungus event occurred after 1945, the nuclear theory would probably have been the first and easy explanation. A piece of nuclear matter had reached criticality on penetrating the Earth's atmosphere.

Much the same arguments are presented to 'prove' this natural nuclear theory as are advanced to account for the accidental explosion of a nuclear-propelled spacecraft. It was not difficult to link the 1908 phenomena to the devastation caused at Hiroshima in 1945, when a comparatively small bomb totally destroyed buildings within half a mile of the detonation point, severely damaged those within one mile and devastated 18 square miles. As in the Tunguska, where the reindeer had developed scabs, Japanese skins had been affected.

But can spontaneous nuclear reaction occur?

Nuclear energy can be released in two ways, by fission and fusion. Fission is achieved by splitting the heavy nuclei of uranium 235 or 238, or plutonium 239. U235 is less abundant than 238 but more readily fissionable. Plutonium is made artificially. The fissionable material must be of correct critical size. If it is too small, there is no fission and no explosion. If too large, it is self-destructive. It melts and makes an inefficient bomb. In order to produce an explosion the material must be made supercritical in time so short as to preclude melting. Several pieces of uranium or plutonium, each smaller than critical size, are brought together very quickly to form one piece. The neutrons then start the chain reaction.

Nuclear fusion, three times more powerful than fission, is brought about by the merging, or fusing together, of light weight nuclei at temperatures of millions of degrees. Sufficient amounts of deuterium and possibly tritium must be contained in a compressed state and triggered by a fission charge. Self-heating carries the reaction to explosion stage. These fission-fusion devices are called thermo-nuclear weapons in which the heavy outer casing of the bomb compresses and reflects back neutrons into the fissionable mass.

Whereas fusion occurs naturally and continuously in the sun and stars, fission has only been achieved, within our present knowledge, by man, and then only with great difficulty.

In nuclear and thermo-nuclear explosions, the materials are converted into hot, compressed gases which expand rapidly, exerting enormous pressures on the surrounding medium and initiating a complex series of events. Light and heat produce thermal radiation and shock waves propagate outwards, rendering the air luminous and creating a fireball. Harmful rays are dispersed. The blast waves cause the initial destruction and main damage. The moving wall of compressed air, generated in a fraction of a second, creates overpressure and drag, the suction wave doubling the power of the pressure wave. The intense heat of the thermal radiation burns, scorches and vaporizes, and the fallout, the radio-active rays, may cause sickness and death. Combined together these forces generate high winds which feed the flames. The damage caused will depend upon the power of the explosion and whether it is detonated on the ground or in the air.

No known nuclear or thermo-nuclear bomb or test device exactly corresponds to the Tunguska explosion which may have yielded energy equivalent to 30 megatons. The two nuclear bombs detonated 1850 feet above Hiroshima and Nagasaki generated a mere 20 kilotons. A ten-megaton bomb is 500 times more powerful. Precise details of the effects of multi-megaton test devices are lacking. The duration of the fireball resulting from a 20-megaton explosion may be 25 seconds and to an observer 50 miles away it may appear brighter than the sun. Its glare may be visible for 400 miles. The radiation flash and fallout may extend over several hundred miles downwind and cover an area of 93,000 square miles, causing skin burns up to 20 miles. The pressure wave can cause damage up to 25 miles.

These rough approximations appear to match in some ways the results observed in the Tunguska. But one essential ingredient of nuclear explosions appears to have been absent. No firm evidence had been found of radio-active fallout, despite somewhat exaggerated claims. But in 1969 something curious was observed.

Knowing by then some of the peculiar effects of nuclear explosions, and recalling possibly the luxuriant plant growth that reappeared on the remnant of the island of Krakatoa after the eruption of 1883, a group of scientists from the University of Tomsk organized a 'floral' expedition to the site.

They found that the fresh growth of trees and plants which had germinated since the explosion, and which normally would have reached 7–8 m (23 to 24 ft), had attained the prodigious height of 17–22 m (about 55 feet), the likely growth after two to three hundred years. The girth of the trees was four times the normal and the rings which, before 1908, had been 0.42 mm thick, were now 5-10 mm. One hundred sections were taken from 100 samples and examined at the Volga Geographical Institute. They showed increased radio-activity after 1908 and the presence of the isotope cesium 137, a radio-active trace element.

What had caused this speed-up of mutation and dramatic change in the plant life within the stricken area? The stunted and wiry trees of 1908 had become a mighty coniferous forest where the natural genetic change had increased a hundredfold.

This prodigious plant growth was puzzling, due to the complete absence of any particle of the body which had exploded above the forest. The shower of tiny particles, a few thousandths of an inch across, had been carried away and distributed in the cloud of dust. To find them seemed a hopeless task. Yuri Lvov found an answer. He knew that peat swamps, such as those within the stricken zone, receive minerals from the air. Thus the moss could form a unique museum of the particles that had fallen from space. Moss grows at uniform rate, making it possible to work down through the layers to any particular point in time.

Lvov found silicate particles within the layer calculated for 1908. Microchemical analysis confirmed that their composition differed from those found in meteor fragments and had no equivalent in known earthly bodies. They resembled the particles contained in the cloud of dust which had caused the bright nights and glaring dawns of 1908. Lvov and his colleagues were to return to this site in the summer of 1976. The full results of their work remain unknown and uncertain. It can be accepted only that there has been remarkable plant growth, whatever may be its cause.

Can this sudden mutation have been due to radioactivity? N. P. Dubinin (*Problems of Radiation Genetics*, 1964) says that under 'conditions of increased radioactivity, wild species are subject to severe selection', which increases heredity changes, some beneficial, others harmful. Ionizing radiation, says C. Auerback (*Genetics in the Atomic Age*, 1956), penetrates living and non-living things and may damage and even sterilize cells. Mutation is likely to occur, and a lethal gene may carry on through successive generations and eventually cause genetic death.

Other authorities on plant genetics and radioactivity (whom I have consulted) reject belief that this speed-up of growth can have resulted from radio-activity. While they agree that, so far, tests have been made only with low doses, they think that radio-activity would be more likely to stunt growth. The reported increase in growth, they suggest, may have been due to the destruction of existing vegetation which allowed the invasion and colonization of species not normally able to compete with the indigenous species. Alternatively, the heat generated by the explosion may have sterilized the soil, giving greater impetus to growth.

This questionable plant growth cannot be taken as evidence that the Tunguska explosion was nuclear. That seems improbable for other reasons which are expanded in the next chapter. The theory envisages a natural, spontaneous release of the nuclear energy contained in a piece of matter which penetrated the Earth's atmosphere. While nuclear fusion occurs naturally but only by the creation of enormous temperatures, spontaneous nuclear fission, within our present knowledge, seems unattainable and inconceivable. The coincidental factors appear to be beyond the realms of reality. But spontaneous fission remains theoretically possible, for in 1940 the Soviet physicists G. N. Flerov and N. A. Petrjak reported that the heavy isotope of Uranium 238 sometimes fissions spontaneously without the aid of any particle.

There may be another way in which nuclei may have been yielded.

14

A Douglas fir tree in Arizona

Professor Willard F. Libby is world famous as the discoverer of the 'Carbon-14' method of archaeological dating, a system made possible by the development of nuclear physics. The basic principles are simple.

The earth is continually bombarded by cosmic radiation, by small sub-atomic high energy particles. They produce small quantities of radio-carbon in the atmosphere, including carbon-14, a rare isotope. This is absorbed by trees and plants and by the animals which eat the plants. The plants and animals retain their carbon-14 in the same proportion as it exists in the atmosphere. When the plant or animal dies and ceases to absorb carbon, the carbon decays spontaneously, slowly and at a constant and known rate. What remains in the sample can be exactly measured and dated. This system rested on the assumption that the concentration of carbon-14 in the atmosphere had remained constant throughout all time. This, however, proved to be untrue for some 6000 years ago it was higher than it is now. That is unimportant as far as the event of 1908 is concerned.

Pursuing his search for ancient trees which might yield additional evidence by their annually formed rings, Libby came across a 300-year-old Douglas fir, the 'Hitchcock' tree as it is known, which had fallen in the Santo Catalina Mountains, about 30 miles from Tucson, Arizona. The wood had been stripped from each ring for the interval of 1870–1936 by the Laboratory for Tree Ring Research of the University of Arizona. The rings for each year were measured for radio carbon. Additional tree rings were measured from an oak tree which had been cut in 1964 in the Simi Valley near Los Angeles. Some 90,000 counts were taken on each sample. They showed a standard deviation of radiocarbon absorption between 1893 and 1928 of 0·005. Only in 1909 had that value been exceeded. The count showed a 1·0 increase. This sudden fluctuation was strange.

Libby discussed the matter with Professor Clyde Cowan, then of the Department of Physics at the Catholic University of America and previously at the Los Alamos Nuclear Research Laboratory, and his own colleague at the University of California, C. R. Atluri. It was not difficult in 1965, by which time the Russian investigation had become well known, to link this marked increase of radio carbon in the atmosphere with the Tunguska catastrophe of 1908.

The three professors wrote an article which was published in the scientific periodical *Nature* on 29 May 1965 (vol 206, pp. 861–865). The article is entitled 'Possible Anti-Matter Content of the Tunguska Meteor of 1908'. The authors do not mean an ordinary 'meteorite', but rather an 'anti-rock'.

First they reviewed the data collected since 1908, remarking certain salient clues. The object had been seen and heard over an area about 1500 km (1000 miles) in diameter. Its explosion had been so blindingly bright and it 'made even the light of the sun appear dark', and it had been accompanied by exceptionally violent radiation and shock phenomena. Seismic, meteorological and magnetic disturbances were registered around the world. The total energy yield was estimated at 10^{23} or even 10^{24} ergs.

There thus appeared to have been a nuclear explosion of some sort yielding neutrons which had contaminated the atmosphere by creating radio carbon.

The experiences of Semenov and Kosotapov at Vanovara were particularly interesting for, having felt burning and seen the sky cleft in two, Semenov just managed to lower his eyes and when he looked again the fire-ball had gone.

Kosotapov, when he heard the thunderous roar, disappearing to the north, ran from his hut but saw nothing. The fire-ball had not lasted long but it emitted so much heat that they could not stand it. Nearer to the disaster zone, Vasiley Okhchen's silverware and samovar had been melted and at the scene 40,000 trees had been knocked down and others scorched and burned.

The radiation flash appeared to be the vital clue to the nature of the object and its explosion. The damage caused and the huge energy yield ruled out a chemical explosion and favoured a nuclear one. By the time these authors wrote it had been established that, in the case of a nuclear explosion, and a fraction of a second after the detonation, a high pressure wave, an intensely hot and luminous shock front forms and moves outwards from the fire-ball.

But, even so, the authors found it hard to believe that the explosion had resulted from either a fission or a fusion chain reaction. In the case of an explosion caused by nuclear fission, it meant that an almost critical mass of fissionable material had become 'tamped', meaning compressed, on entering the atmosphere, carrying the material far beyond criticality very quickly to prevent its disassembly with low yield. (Such a process has been described as trying to set on fire a pile of wet leaves.) Even so, had criticality and fission resulted, the observed multi-megaton yield would have required large initial mass, well above the critical mass of normal density uranium or plutonium. Such supercriticality obtained by tamping could hardly be credited as the mechanism of the explosion.

Explosion from fusion reaction seemed even more improbable. For that to have been achieved a sufficient amount of deuterium and possibly tritium must have been contained in a compressed state and heated to several million degrees centigrade. It must be maintained in that state so that self-heating can carry the reaction to explosion stage. Such a mechanism could not have been attained merely by entry into the Earth's atmosphere.

The authors searched for other natural means by which to account for the release of nuclear energy – undoubtedly, in their opinion, the cause of the explosion. They could find only the annihilation of matter by anti-matter.

The theoretical existence of anti-matter was proposed in 1930 by the English mathematician and theoretical physicist, Paul Adrien Maurice Dirac. Dirac reasoned, because each particle should have an 'anti-particle', that there must be an 'anti-electron' exactly like the electron except that it had a positive instead of a negative charge, and an 'anti-proton' with a negative instead of a positive charge. Dirac's anti-electron was found in 1932 by the American physicist Carl David Anderson working with Robert Millikan at Cal Tec. They observed on a photographic plate a track that curved in the wrong direction. It had the same mass but opposite charge. There it was – Dirac's 'anti-electron'. Anderson named it the 'positron'. It has only an instant of existence. It vanishes for, when it encounters an electron, the two annihilate each other, creating pure energy in the form of gamma rays. This was startling confirmation of Einstein's theory that matter could be converted into energy and vice versa – the suggestion that led eventually to the birth of the bomb. Thus, all three particles which make up matter, the proton, neutron and electron had anti-particles. But each has a different spin, one is left and the other right. Libby, Cowan and Atluri agreed that there was no evidence for the existence of anti-matter other than theoretical. No piece of anti-matter had been observed, unless, according to a recent theory, ball lightning results from the collision of matter and anti-matter.

Another problem was the flight of a rock formed of anti-matter through our atmosphere. How far could it penetrate before being annihilated by collision with matter? That would depend on the rock's density. It would have thinned down and become consumed by absorption with the gases of the atmosphere. To have reached the required distance and to have yielded energy to the order of 10^{24} ergs, the rock would have needed a diameter of 100 cm (39½ ins).

The authors' first thought was that, even so, it should have been consumed earlier and have exhibited its largest yield somewhere towards the middle of its flight-path rather than towards its end. But a second look at the facts modified this conclusion.

The atmosphere ahead of the 'rock' would have become rarefied by the exceedingly strong accompanying radiation shock together with the forward ejection of complex anti-nuclei. That would have heated the air and greatly increased the object's range. Only a small fraction of the anti-matter rock would have become annihilated in flight. It would remain essentially solid until it reached the point deep in the atmosphere where continued annihilation heated it into a gaseous state, causing it to disintegrate explosively.

The neutrons released were rapidly oxidized into carbon dioxide in the atmosphere, some reaching Arizona where they were absorbed by the ancient Douglas fir tree. The authors estimated that there had been an increase of radio-carbon activity in the atmosphere of some 7 per cent. They based their estimate on both the uniform rate of distribution and absorption, and on the yield of radiocarbon deriving from the testing of nuclear explosions in the atmosphere. By September 1961 the equivalent of 70 mt of fission and fusion nuclear explosives had been released in air-bursts, and about 100 mt in surface tests. Since July 1945 plant life had absorbed radiation at 25 per cent above the natural cosmic-ray level, one half being retained and the other decaying at known rate.

When the scientific detectives used the known damage figures of the Tunguska explosion as input data for the Nuclear Bomb Effects Computer a value of about 30 mt (10^{24} ergs) energy yield was obtained per neutron pair consumed. That gave a total neutron yield of $(2\cdot7\pm1\cdot4)+10^{27}$ neutrons. Since the rock had disintegrated in the atmosphere it would be expected to give $(2\cdot7\pm1\cdot4)+10^{24}$ carbon-14 atoms.

Thus, had the explosion been due to anti-matter it would have been equivalent to 35 mt of fission or fusion.

Cowan's, Libby's and Atluri's hypothesis that the Tunguska explosion had been caused by the annihilation of matter by anti-matter did not pass unchallenged. Writing in *Nature* (vol 211, 3 September 1966) Robert Gentry, of Columbia College, Takoma Park, Maryland, pointed out that the observations of Semenov and Kosotapov showed that the fire-ball did not last long, presumably only for a few seconds. That was difficult to reconcile with the usual 33 second fire-ball to be expected from a 30 megaton thermo-nuclear explosion, the calculated equivalent yield of the Tunguska burst. The relatively short duration of the fire-ball appeared to be reasonable evidence against the Tunguska explosion being thermo-nuclear in nature.

L. Marshall of the University of Colorado supported Cowan, Libby and Atluri (*Nature*, vol 212, 10 December 1966) and re-affirmed the validity of their conclusions.

J. C. Lerman, W. G. Mook, and J. C. Vogel of the C-14 Research Unit, Physics Laboratory, University of Gröningen in the Netherlands questioned Libby's sampling of the tree rings (*Nature*, vol 216, 9 December 1967). New measurements, they suggested, supplied higher accuracy and the result showed no significant deviations which could be connected either with the Tunguska explosion or with the sunspot cycle. Eleven samples had been taken from trees growing near Tronheim in Norway. They may have been felled in 1957. Cowan and his collaborators had estimated an increase of seven per cent for the global atmospheric content of carbon-14 as a consequence of the interaction of a mass of anti-matter large enough to produce the energy released by the Tunguska explosion, an increase of about one per cent of carbon-14 in 1909. The Dutchman, however, concluded that no deviation larger than three per mil had been measured and that any possible deviation around 1909 must have been small.

As the principal advocate of the comet theory Vasily Fesenkov discussed the new hypothesis, writing in the *Journal of the British Astronomical Association* (78·2) in 1968. Quoting a number of Soviet scientists including V. I. Baranov and F. I.

Pavlotsky, he pointed out that there had been no rise in radio-activity at the spot where the Tunguska body fell. Therefore it had nothing in common with anti-matter for the annihilation of matter must be attended by radiation.

But, assuming that the explosion was caused by the annihilation of matter by anti-matter, from whence came the anti-rock? According to one theory our universe is composed of matter and anti-matter. Both exist in our galaxy, and their mutual annihilation may account for the enormous energy and brilliance shown by quasars, the enigmatic objects perceived to be hurtling outwards at near the speed of light in the extreme depths of our universe. These quasars may even be composed of colliding galaxies, one formed of matter, the other of anti-matter.

According to another theory, at the time of the original Cosmic Egg, the primordial Big Bang, both matter and anti-matter were created in equal parts. Hurled apart by propulsion they split into two universes, the anti-matter universe becoming the mirror-image of our own familiar world. This possibility was forecast by Dirac, who was awarded a Nobel Prize in 1933.

If the Tunguska phenomenon resulted from the annihilation of matter by anti-matter, the anti-rock may have come from within our universe or from another universe. Its possible existence will come up for discussion in a later chapter when an even more startling hypothesis for the Tunguska event will be presented.

15

Cosmological theories

'It' may have been a tiny 'black hole', one that had existed since time began. That solution is suggested by A. A. Jackson IV, and Michael P. Ryan, Jnr, of the Center for Relativity Theory, University of Texas (*Nature*, vol 245, 14 September 1973).

A black hole, as its name implies, is a region of space from which nothing, not matter nor any signal, nor even light itself, can escape. Its gravitational force–its pull–the escape velocity, is greater than the speed of light. It is empty because anything attracted within it may be crushed out of existence. It swallows up everything that comes in its path. As it drags in more matter its mass increases, and so does its power of attraction. It sweeps up everything like a maelstrom in space. It has been called a 'celestial vacuum cleaner'. Matter attracted within it reaches incredible densities and simply vanishes. The long-accepted laws of nature cease to apply.

The possible existence of such a bizarre phenomenon was predicted as long ago as 1798 by the French mathematician Pierre Laplace. He calculated that if a celestial body was sufficiently dense or massive, thus having enormous gravitational attraction, it would be invisible because light would be unable to travel fast enough to escape from its surface.

Albert Einstein in 1905 also predicted the existence of what we now call 'black holes'. Explaining gravity in terms of 'space-time' curvature, he suggested that such phenomena should occur at the end point of stellar evolution. Modern physicists agree that if black holes do not exist, it would be necessary to revise the Special Theory of Relativity upon which all physical concepts are based.

The situation is similar to that posed by the eighteenth century philosopher Voltaire who suggested that if God did not exist it would be necessary to invent Him. And, like God, black holes exist only in theory. No one has, or ever will, 'see' one. But if they do not exist all theories of the origin, evolution and ultimate fate of our universe would cease to apply.

Hundreds, thousands, and even millions of black holes may be present in our universe. A large one may be at work at the centre of our galaxy slowly devouring all its matter, its 100 billion stars including the sun. Our universe may itself be a huge black hole, a vortex into which everything, all the billions and billions of stars and galaxies, are being sucked and may be crushed out of existence, or will simply disappear.

The discovery of black holes of the 1960s is described as the most exciting achievement of physical science, for it has created an entirely fresh concept of the cosmos, one that is foreign to western philosophy but possibly closer to eastern ideas. We westerners have been schooled to believe in an inevitable progress of events from Creation to Ultimate End, whereas Asiatic philosophy conceives of no Creation or Creator, and no finite end.

What goes on in the eastern mind, in the words of Professor John Wheeler of Princeton University, one of the founders of the new cosmology, is 'a process in which all shapes are simultaneously present', and is 'a set of logical not of chronological relationships'. He cites as an example the story told by an acquaintance who travelled through one of the great nations of South-East Asia in 1952. Observing how many children were dying whose life might have been saved by small-pox vaccination, he could not forbear to express his dismay to the Minister of Education. That official could not have been less concerned. 'After all what difference does it make; they will be reincarnated anyway.'

We need to revise many of our ideas to understand the new cosmology. Space is no longer the orderly concept of Euclid and Newton. It is even more complex than Einstein's Space-Time Curvature Equation. It is weirder than we can imagine.

And it is no longer true to say that the existence of black holes is solely theoretical. Their presence can be inferred by the peculiar behaviour of their near neighbours and from the powerful radiation some disperse.

The creation of a black hole is the logical consequence of stellar evolution. Within our universe stars are continually born and die. These massive glowing 'suns' are formed by gravitational forces which attract and condense the gases which float in space. They radiate energy in the form of heat, light and other rays. They do that for millions of years until their nuclear fuel becomes exhausted.

A star, when it has burned up its fuel, becomes transformed beyond recognition. How it behaves will depend upon its original mass.

An average-sized star, having about the mass of the sun, after shining brightly for about 7 million years will swell to an enormous size becoming a 'red giant' some 200 million miles in diameter. As it burns up more of its fuel it becomes even more bloated. Expansion halts and it contracts below its normal size, becoming smaller and smaller until it stabilizes as a 'white dwarf', about the size of the earth. Further contraction is impossible because all its electrons have become closely packed together. Its atoms are squashed by tremendous gravitation. Its density is so great a matchbox full would weigh several tons. There are many such white dwarfs in our galaxy.

More massive suns follow a different evolutionary path. A star of two sun masses, and many are 50 times more massive, will expand to enormous size and recontract. But it will be unable to settle down to equilibrium as a white dwarf. It bursts asunder, blowing off its outer surface in catastrophic explosion and releasing tremendous energy.

The last Super Nova explosion seen in our galaxy was in 1604 and was observed by the astronomer Johannes Kepler. Even more famous is the catastrophic explosion of the Crab Nebula in 1054 which was watched by the Chinese astronomers. Due to its distance of 4000 light years from the earth, the time in which its light took to reach us, what the astronomers saw had happened about 3000 years before Christ.

All evolving star masses reach the white dwarf stage. They can only find equilibrium by contracting even further to neutron stars, having a density 100 times greater than a white dwarf so that a match-box full would now weigh the equivalent of an asteroid a mile in diameter. It may have a radius of only 10 km (6 miles) and is a sort of overgrown atomic nucleus but bound together by gravitational rather than by nuclear forces.

The existence of these tiny neutron stars was predicted in 1932 by the Soviet physicist Lev Landau and confirmed theoretically in 1939 by the American nuclear physicist Robert Oppenheimer and others. They were discovered in 1967 by the radio and optical pulses emanating from the collapsed remnant left behind when the Crab Nebula exploded.

The forces of gravitation within the neutron star overwhelm the forces resisting compression. The linear contraction from red giant to neutron star may be 20,000,000 to 1. Its evolutionary process is not yet complete.

Were we able to watch it through a giant telescope

we would see, like the grin of the Cheshire Cat, its light fading and becoming dimmer and dimmer as it collapses even further, leaving behind its grin and becoming lost to view in the form of a black hole.

There are two kinds of black holes depending upon the shape of the original star. Circular stars become non-rotating black holes with a spherical surface. They are named the 'Absolute Event Horizon', the zone where the density and powerful gravitational pull prevents the escape of light or anything that falls within. Nothing can get out. Everything is presumably crushed out of existence by the tremendous tidal forces of gravity. As far as our universe is concerned all matter attracted within the event horizon must disappear completely and for ever.

Time as we understand it ceases to operate for the event horizon is an absolute barrier to any communication within or without. This confusing situation is best explained by the analogy of the unwary astronaut who becomes sucked into a black hole, an unlikely possibility for they are mere pinpricks in the vast emptiness of space. We need also to imagine that we can watch his progress as he reaches the boundary of the event horizon, the Schwarzchild radius as it is called after the German mathematician who in 1917 calculated the critical size of the time-trap within a black hole.

We, the observers, would see the astronaut moving ever more slowly, becoming fainter and fainter until he appears to stand still, his image frozen for ever. He appears to hang motionless like a dead fly on a spider's web.

But the astronaut himself will notice nothing particularly unusual as he crosses the event horizon and enters within a few thousandths of a second the Space-Time Singularity. The event horizon is the boundary to the unknown. Within it concepts of space and time become interchanged. Viewed from outside time marches on. It is the reverse inside. If the astronaut looked backwards he would see a copy of himself falling inwards. Should he be able to consult his clock he would see its hands reverse and go backwards faster and faster. His body may become elongated, stretched to infinite length, or he may be crushed out of existence, squeezed by infinite density.

Is his body completely crushed or will it cross a bridge into another universe? Either way he can never return to ours. All known laws of physics break down. According to the law of conservation matter can never be destroyed. It merely changes its form.

What happens within a black hole has posed a terrible dilemma which several physicists have sought to answer. According to R. M. Hjellming of the National Radio Astronomy Observatory at Green Bank, Virginia, matter that disappears within a black hole is not crushed out of existence. It is recreated in another universe also connected to ours by a white hole, a black hole in reverse. Thus matter, or matter and anti-matter, are interchangeable between multi-connected universes. There is nothing in the General Theory of Relativity that says that cannot occur. In fact it may have happened at the time of the primordial Big Bang, more than 12 billion years ago, the initial convulsion which created our universe, possibly from matter deriving from another universe. And our universe may itself be a black hole, sucking in and ejecting all its matter into another universe. It may be significant that the estimated radius of our universe conforms to the Schwarzchild Radius for the width of an event horizon.

This awesome possibility seems to be confirmed by the experiments conducted by Joseph Weber of the Princeton Institute for Advanced Study. He placed two instruments 600 miles apart at Chicago and in Maryland to detect waves of gravitational energy sweeping in from the denser parts of our galaxy. They came in violent bursts, apparently from nowhere, and were never repeated, unlike what might be expected from large masses. They could have been produced only by events of tremendous violence. Entire stars were being annihilated, vanishing completely into black holes. And these events are occurring daily.

But, and therein lies the impossible paradox, if Weber's results are correct our universe must long ago have run out of matter for our galaxy for example contains 100 billion stars which are 10 billion years old. If one star vanishes each day and has been doing so for that number of years, our galaxy could only have existed for 270 million years. Life would not have had time to develop on our earth which came into planetary existence 4·5 billion years ago.

Here Hjellming steps in again. All physical laws demand that as fast as matter disappears from our universe, it must re-emerge from another universe, through white holes.

According to the black hole-white hole theory our universe will expand without limit till expansion is halted by gravitational forces. It will then retract, collapsing into a black hole to be reborn in another universe.

Alternatively, expansion eventually ceases and all the stars and galaxies are brought together again in one mass which explodes in yet another Big Bang. In this theory the process of explosion, expansion, contraction, explosion, continues for ever.

Either way, our universe is still expanding for the further we gaze into space the faster the galaxies are receding. But according to calculations made at Cal Tec this expansion is beginning to slow down. It will stop and retraction will begin in about 70 million years.

Why this may be happening is a mystery because our universe contains less than one-tenth of the mass required to exert the gravitational pull to halt expansion. John Wheeler has called it the 'case of the Missing Matter'. It has disappeared, he believes, into super-space, the everlasting physical background to our universe, to all universes, past, present or future.

Trying to define super-space is described by Wheeler as 'like chasing after Merlin. One moment it is a rabbit and the next a gazelle. And just as you reach out to touch it, it turns into a fox, or a brightly-coloured bird fluttering on your shoulder. It is the place where smoke comes out of the computer because all the classical laws of space-time break down.' He conceives super-space as a region in to which all the planets, stars and galaxies would eventually disappear, swirling down a time-tunnel like water gurgling out of a bath. In super-space time stands still and the events of a billion years are compressed into a split second. Concepts of time like 'before', 'after' and 'next' lose all meaning.

Super-space, in Wheeler's view, is connected to our universe and to other universes by 'worm-holes' which provide access to secret paths through which fantastic journeys can be made in an instant of time by a series of 'jumps' through time barriers. He visualizes it like this: take a sheet of paper and cut two little dots, one representing earth and the other a star many light-years distant. Moving down the paper a signal would take that many light-years to reach from one to another. But if we fold the paper so that the two dots, or 'worm holes', coincide the signal takes no time at all to make the journey. The signal can traverse the whole of super-space or instantaneously connect two points in one universe, or move into another. Space travel might be possible by these worm holes – the space-craft moving instantly from one world to another.

To dismiss the existence of super-space we would have to reject relativity in its entirety and deny the existence of nuclear energy. Einstein in 1935 hinted at the existence of bridges connecting two or more widely separated parts of the universe and he believed it possible that 'other universes exist independently of our own'. His notions seemed so mad that many scientists sought to reject his theories of relativity until Roger Penrose, Professor of Mathematics at the University of Oxford, pointed out that similar phenomena followed from any alternative theories of space-time.

Other scientists have independently reached the same conclusions as Wheeler. Kip Thorne of Cal Tec thought that after a star had disappeared through a black hole its matter might re-emerge, bubbling upwards like a mountain spring, in some other region of the universe. Models of stellar collapse showed that this is precisely what happens. The Israeli scientist Yuval Neeman likened the multiple universes to 'two trouser-legs', the seen and the unseen, connected by tunnels.

Some scientists hope one day to prove the existence of super-space by finding a means whereby to transmit a signal through space by ordinary radio and simultaneously through the secret paths of super-space. If it reaches its goal or returns faster than the speed of light, then the signal can only have travelled through super-space.

Whether or not super-space exists, there is no doubt of the existence of black holes. The secret was disclosed by binary stars where a very massive star is accompanied by an invisible companion. Such a combination was found in connection with the super-giant H D E 26868 which is 6500 light years distant from the earth, and is 30 times more massive than the sun. Every 5½ days it circles in orbit an invisible companion less than half its mass but utterly tiny with a radius of only 30 miles. Despite its small size its gravitational influence distorts the larger star out of spherical shape. It becomes egg-shaped and material is dragged from it, falling inwards towards its companion and is slowly funnelled into its aperture. The companion is named Cygnus X-1 and from it very hot X-rays are emitted. The colossal gravitational influences exerted by

Cygnus X-1 prove it is a black hole.

Another method of detection was suggested by the Soviet astro-physicists, Snklovskii, Zeldovich and Novikov. They theorized that if gas is drawn into a black hole it will become so tremendously heated by compression that it will emit copious quantities of X-rays. This was particularly likely to occur within rotating black holes, those derived from 'rugby-football' shaped stars. The emission of powerful X-rays from Cygnus X-1 was confirmed in 1969 by Joseph Weber.

The likelihood that black holes are sources of vast energy has prompted the thought that if it can be trapped it would be possible to construct a bomb of unbelievable capability and provide unimaginable power. Even more disturbing is the picture of the mad scientist who may try to create a black hole in his laboratory. If it slips from his grasp it might penetrate the earth devouring its substance as it sinks to its centre, causing eventually total gravitational collapse. Whether or not such an experiment is feasible, it is likely that some black holes in space may end up by exploding as violently as a million megaton hydrogen bomb.

Black holes may not last for ever. Professor Stephen Hawking of Cambridge University has suggested that, as they lose radiation from within, they may also lose mass and eventually disappear. The greater their mass, the greater is the rate at which radiation and mass are lost. He envisages also the possibility of there being tiny black holes of very low mass. His suggestion ('Monthly Notes', *Royal Astronomical Society* 152, 75-1 (1971)), stimulated the thoughts of Jackson and Ryan.

16

But no exit?

'There have been many attempts to explain the Tunguska meteorite, ranging from the prosaic to the bizarre', Jackson and Ryan remark at the beginning of their paper. A black hole of substellar mass could possibly explain the mysteries associated with the event. It may have had the mass of a small asteroid, having a strong gravitational field, with a velocity slightly greater than the Earth's escape velocity. Such an object, they suggest, would set up both bow and tail shock-waves. They estimate the total energy in the blast-wave as 10^{22} and 10^{24} ergs. It would have reached temperature peaks between 10^4 and 10^5 K, so most of the radiation from the shock-waves would be in the vacuum ultra violet and would be absorbed at longer wavelengths. There would be little hard 'x' radiation and the accompanying plasma column would appear deep blue.

These estimates, they consider, agreed well with the eye-witnesses' reports and the measurements of the pattern of the throwdown of trees. The witnesses described the object as a bright blue 'tube' and the searing of the trees indicated a temperature compatible with these calculations. The shock-wave from the black hole resembled that produced by Zotkin's and Tsikulin's model.

The black hole would leave no crater or material residue. It would penetrate the earth, where the rigidity of rock would allow no underground shock-wave. Due to its high velocity, and because it would lose only a small fraction of its energy in passing through the earth, the black hole would have followed very nearly a straight line, entering at 30° to the horizon and exiting through the North Atlantic in the region of 40°-50° N, 30°-40° W.

That provided a check for the hypothesis. At the exit point there would have been another shock-wave and disturbance of the surface of the sea. Had there been, the microbarograph and shipping records should show it.

Jackson and Ryan were answered by two writers also from the University of Texas, William H. Beazley and Brian A. Tinsley (*Nature*, vol 250 16 August 1974). They agreed that the air blast could have resulted from the impact of such a small black hole which would have passed through the earth in 10-15 minutes, causing a similar explosion at the point of exit in the North Atlantic.

The English microbarograph records recorded waves coming from Siberia, 5720 km (3500 miles) from the point of impact, travelling at 320 metres per second. They were recorded at approximately 05·15. Shock-waves from the far closer point of exit should have been recorded in London between 2·15 and 3·30, three hours earlier. The microbarographs had not registered such waves. The logs of the British Fleet on manoeuvres in the North Atlantic recorded no unusual sea disturbance. Thus, either Jackson and Ryan are wrong, or (a possibility they do not consider) the black hole remained within the earth devouring and eventually destroying its substance.

Two other writers to *Nature* (vol 247, 18 January 1974), Gerald L. Wick and John D. Isaacs, of the Scripps Institution of Oceanography, La Jolla, California, found Jackson's and Ryan's hypothesis both imaginative and intriguing. But unfortunately, they thought, the miniature black hole could not account for all the important phenomena known to accompany the event, for example the subsequent bright nights and the high nickel content of the magnetic globules found in the stricken area. That data supported the cometary theory. But discussion of the event did not preclude the possibility that a black hole comprised the nucleus of the comet, condensing its material.

In a letter to this author, 15 November 1975, Jackson agreed that the exit pulse had not been found on the microbarographs, so, 'Alas, it seems that the Tunguska event was not caused by a black hole'. He and Ryan thought that the cometary hypothesis was still in dispute. Discussing the other theories, they concluded: 'It begins to seem that the Tunguska event is more bizarre than any explanation put forward to date'.

17

Or, even more bizarre?

No single explanation for the Tunguska mystery is wholly satisfactory, or is accepted by the many scientists who have theorized about the problem. Each investigator pursues his own course of inquiry, and every theory has been contradicted. Research at the site continues, and discussion has become worldwide.

The comet theory has been strongly supported by recent articles in Western scientific journals, particularly by John C. Brown, of the Department of Astronomy, University of Glasgow, and David W. Hughes of the Department of Physics, University of Sheffield, writing in *Nature* (Vol. 268, 11 August 1977). They are supported by Ian Ridpath, the editor of the *Encyclopedia of Astronomy and Space*, writing in the *New Scientist*, 11 August, 1977.

Opposition to the cometary theory, remark Brown and Hughes, and consequent support for the less conventional hypothesis involving black holes, almost-critical masses of extraterrestrial fissionable material, anti-matter bodies and alien spacecraft, have been based substantially on the non-observance of the comet before impact, the explanation of the height of the explosion above the Earth's surface, the composition of the glassy spherules found at Tunguska, and most important, the apparent occurrence of nuclear phenomena in the explosion as indicated by the subsequent enhancement of radio-carbon in the atmosphere. They consider the first and last points to be the most important.

They assume from the data collected that the Tunguska comet was one of the more well-behaved members of the species and obeyed the equations set for such objects. When it encountered the Earth it came from a point in the dawn sky comparatively close to the Sun and would thus have been most difficult to detect and observe. Its orbital configuration was similar to Comet Mrkos which was only detected in 1957 after it had rounded the Sun and travelled beyond the Earth's orbit.

One of the major stumbling blocks of the cometary impact hypothesis is the carbon anomaly that is said to have occurred in 1909. Cowan, Libby and Atluri found that the carbon activity was higher that year than normal, and proposed that this extra radio carbon in the atmosphere was due to anti-matter annihilated at Tunguska. Hunt, Palmer and

Penny attributed this carbon excess to a nuclear explosion.

Other authors have previously remarked that the only evidence for possible nuclear effects derives from the radioactivity found in two trees, one growing near Los Angeles, California, and the other near Tucson, Arizona, while trees growing far nearer the scene of the catastrophe, at Tronheim in Norway, showed no radioactive increase in 1909, but rather a steady decrease. The only radioactivity at the site was dismissed in 1962 by Florensky as the fallout from subsequent atomic bombs.

Brown and Hughes explain this radio carbon data by the physics of a cometary entry into the atmosphere. They note that, although the temperature produced by the entry of a comet into the atmosphere would have been no more than a few million degrees, too low for nuclear reactions, it is "entirely fallacious" to suppose that sub-nuclear temperatures cannot produce nuclear effects. Such temperatures are produced in solar flares which are similar to a comet exploding in the atmosphere. Thus the impact of the comet with the atmosphere would have produced X-rays, gamma rays, and highly accelerated electrons and nuclei. Sufficient neutrons could have been produced to account for the radio-carbon data found by Cowan and his colleagues. Thus, even if the Tunguska event did cause nuclear effects, that would not invalidate its identification with a comet. 'Nothing more exotic need be involved', comment Brown and Hughes.

Taking up their argument, Ridpath remarks that

a smaller version of the Tunguska event occurred on 31 March, 1965 when a meteorite fragmented explosively above the town of Revelstoke, Canada, leaving no crater and depositing traces of black dust which showed that the meteorite had been a carbonaceous chondrite, a composition typical of interplanetary debris, including comet heads. Carbonaceous chondrites are so fragile that few survive the passage through the atmosphere.

Dealing with the skeptics who have doubted that an object large enough to have caused the devastation in the Tunguska would have been invisible, Ridpath points to a practical demonstration which occurred in 1976 when an asteroid passed the Earth at a distance of just over a million KM, avoiding a direct repetition of the Tunguska event by only a few hours. Although termed an asteroid, it was one of that class of objects believed to be the nuclei of a "dead" or degassed comet. Its diameter was similar to the Tunguska comet. Even at its closest it was too faint to be seen without a large telescope, and at the same closing speed as that calculated for the Tunguska object, 40 KM/s, would not have been visible to the naked eye until 25 minutes before impact had it come out of a perfectly dark sky. But, of course, the Tunguska object struck in daylight. It is not surprising, therefore, that the Tunguska comet was not seen as it approached the Earth on that sunny morning.

The investigations carried out at the site by the Soviet Academician G. I. Petrov and his colleagues from the Tomsk University each summer have established that as the cosmic body moved through the atmosphere its substance rapidly evaporated and, when a large amount of vapour had amassed in front of the travelling body, it exploded and scattered in the atmosphere, condensing into tiny balls which gradually descended and settled over a vast area. Small fused silicate particles were found in the 1908 peat layers, and microchemical analysis showed that they were, surprisingly, composed of a set of elements most unusual in classical meteorites—they were rich in rare earth and heavy elements. Similar 'Tungus' elements were found during rocket probes of noctilucent clouds. But the quantity of meteorite particles found over the area of the felled trees proved to be very small, no more than several tons, while the total mass of the object had been estimated at 100,000 tons. What had happened to the greater proportion of the cosmic matter?

Laboratory analysis provided the answer, according to Professor Vasiliev. "Numerous soil and peat samples were taken last summer (1976) in the area of the epicentre to determine their cosmogenic radiocarbon content. An analysis of the samples at the Institute of Geochemistry and Mineral Physics, Ukranian Academy of Sciences, revealed that a considerable amount of cosmogenic material had fallen near the epicentre in the form of silicate particles. It is quite possible that the total amount within the area of destruction alone runs into at least thousands of tons.' The scientists carried out the delicate job of detecting this cosmic dust in the 1908 peat layers. This proved that the silicate particles, the product of smelting at the moment of explosion, were of original and extraterrestrial origin. This latest data, found the Tomsk scientists, made it likely that the Tunguska object was a comet, as did its lengthy tail which accounts better than any other hypothesis for the cloud of dust which permeated the atmosphere for several days. But it leaves the intensity of the explosion unexplained, unless the comet had enormous velocity and huge mass.

The likelihood that the Tunguska object was a comet is not as reassuring as it sounds, in comparison with nuclear matter, anti-matter, black holes or crashing space-craft, because another comet could strike the Earth, as one nearly did on 10 August 1972. Observed at low trajectory over the western United States it ricochetted off the atmosphere, disappearing into space. Had it struck with the explosive force of a thermo-nuclear bomb at such a time of international tensions, it might have sparked off a nuclear war before its true nature could have been determined.

But maybe the Tunguska event is more bizarre than we can imagine. The intensity of astro- and micro-physical research is continually opening up fresh fields of inquiry which may one day provide a completely satisfactory answer. At least it has been spared the explanation that it was due to some relic of lost Atlantis, or encompassed within the ever expanding Bermuda Triangle.

Bibliography

(In English Language)

Astapovitsch, J. S., 'Air Waves Caused By the Fall of the Meteorite on 30 June 1908', *Quarterly Journal of the Royal Meteorological Society,* 60, 493-504 (1934).

Beazley, W. H. and Tinsley, B. A., 'Tungus Event Was Not Caused by a Black Hole', *Nature*, vol 250 (16 August 1974).

Ben-Menahem, Ari, 'Some Parameters of the Siberian Explosion of 30 June 1908, From Analysis and Synthesis of Seismic Signals at Four Stations', *Physics of the Earth and Planetary Interiors,* 11, 1-35 (1975).

Cowan, C., Atluri, C. R., Libby, W. F., 'Possible Anti-Matter Content of the Tunguska Meteor of 1908', *Nature*, vol 206 (29 May 1965).

Fesenkov, V. G., 'On the Cometary Nature of the

Tunguska Meteorite', *Soviet Astronomy*, A.J. vol 5, no 4 (January–February 1962).

Fesenkov, V. G., 'A Study of the Tunguska Meteorite Fall', *Soviet Astronomy*, A.J. vol 10, no 2 (September–October 1966).

Florensky, K. P., 'Cosmic Dust and The Present State of the Investigation of the Tunguska Meteorite', *Geochemistry*, no 3 (1963).

Glasstone, S., 'Effects of Nuclear Weapons' (US Government Printing Office, Washington, DC, 1960).

Hughes, D. W., 'Tunguska Revisited', *Nature*, vol 259 (26 February 1976).

Hunt, J. N., Palmer, R., Penny, Sir William, 'Atmospheric Waves Caused by Large Explosions', *Atomic Energy Authority* (1959).

Ivanov, K. G., 'The Height of the Explosion of the Tunguska Meteorite', *Soviet Astronomy*, A.J. vol 7, no 2 (September–October 1963).

Jackson, A. A. and Ryan, M. P., 'Was the Tungus Event due to a Black Hole?', *Nature*, vol 245 (14 September 1973).

Krinov, E. L. and Romankiewicq, J. S., 'Giant Meteorites', (Pergamon, Oxford, 1966).

Krinov, E. L., 'The Tunguska and Sikhote-Alin Meteorites', in *The Solar System*, by Middlehurst and Kuiper (Chicago, 1963).

Nininger, H. H., *Our Stone Pelted Planet*, (1953).

Parry, A., 'The Tungus Mystery. Was it a Spaceship?' in *Russian Rockets and Missiles* (Macmillan, Cowan, 1960).

Penrose, R., 'Black Holes' in *Cosmology Now* (BBC, 1973).

Wasson, J. T., *Meteorites – Classification and Properties* (Springer-Verlag 1974).

Whipple, F. J. W., 'The Great Siberian Meteorite and the Waves, Seismic and Aerial which it produced', *Quarterly Journal of the Royal Meteorological Society*, 56, 287–304 (1930).

Whipple, F. J. W., 'On the Phenomena Relating to the Great Siberian Meteorite', *Quarterly Journal of the Royal Meteorological Society*, 60, 505–513 (1934).

Wick, E. L. and Isaacs, J. D., 'Tungus Event Revisited', *Nature*, vol 247, no 4537 (18 January 1974).

Yarvnel, A. A., 'The Composition of the Tunguska Meteorite', *Geochemistry*, no 6 (1957).

Zolotov, A. V., 'Estimation of the Parameters of the Tungus Meteorite Based on New Data', *Soviet Physics–Doklady,* vol 12, no 2 (August 1967).

Zolotov, A. V., 'The Possibility of "Thermal" Explosion and the Structure of the Tungus Meteorite', *Soviet Physics–Doklady*, vol 12, no 2 (August, 1967).

Zotkin, I. T. and Tsikulin, M. A., 'Simulation of the Explosion of the Tunguska Meteorite', *Soviet Physics–Doklady*, vol 11, no 3 (September 1966).

The periodicals listed are available at the Science Reference Library, London, which also possesses many articles in Russian. I am authoritatively informed that all important papers dealing with the Tunguska problem have been translated.

The Investigators

C. R. Atluri, Department of Physics, Catholic University of America.

Ari Ben-Menahem, The Adolpho Boch Geophysical Observatory, Department of Applied Mathematics, The Weizman Institute of Science, Rehovot, Israel.

C. Cowan, Department of Physics, Catholic University of America.

V. G. Fesenkov, Astrophysical Institute, Kazakh Academy of Sciences, Member of the Committee on Meteorites, Academy of Sciences, USSR.

K. P. Florensky, V. I. Vernadskii Institute of Geochemistry and Analytical Chemistry, Member of the Committee on Meteorites, Academy of Sciences, USSR.

K. G. Ivanov, Institute of Terrestrial Magnetism,

Ionosphere and Profagation of Radio Waves, Academy of Sciences, USSR.

E. L. Krinov, Secretary of Committee on Meteorites, Academy of Sciences, USSR.

W. F. Libby, Founder of Radio-Carbon Dating Technique, Department of Chemistry, University of California, Los Angeles.

A. A. Yarvnel, Committee on Meteorites, Academy of Sciences, USSR.

A. V. Zolotov, All Union Scientific-Research Institute of Geophysical Prospecting Methods, Volga-Ural Branch, and A. F. Joffe Physicotechnical Institute, Academy of Sciences, USSR.

I. T. Zotkin and M. A. Tsikulin, O. Yu Schmidt Institute of Geophysics, Academy of Sciences, USSR. Members of Meteorite Committee, Academy of Sciences, USSR.

About the Author

Rupert Furneaux, who was educated at Eastbourne College, has been a professional writer for more than twenty-five years. In that time, he has published sixty books and has established an international reputation for his many distinguished works of non-fiction. Before taking up a career as a full-time writer, Rupert Furneaux worked in the film industry, directing and producing documentaries.

An intrepid researcher, he has travelled throughout the Middle East, Africa, the United States, the Bahamas and the West Indies collecting evidence and information on the world's great mysteries and natural disasters. Among his bestselling titles in this field are *Krakatoa: The Great Eruption of 1883* and *The Money Pit Mystery*.